What little I remember

What little I remember

OTTO R. FRISCH F.R.S.

CAMBRIDGE UNIVERSITY PRESS

CAMBRIDGE
LONDON · NEW YORK · MELBOURNE

Published by the Syndics of the Cambridge University Press
The Pitt Building, Trumpington Street, Cambridge CB2 1RP
Bentley House, 200 Euston Road, London NW1 2DB
32 East 57th Street, New York, NY 10022, USA
296 Beaconsfield Parade, Middle Park, Melbourne 3206, Australia

First published 1979

Printed in Great Britain at the University Press, Cambridge

Library of Congress Cataloguing in Publication Data
Frisch, Otto Robert, 1904–
What little I remember.
Includes index.
1. Frisch, Otto Robert, 1904– 2. Physicists –
Great Britain – Biography. I. Title.
QC16.F75A38   530'.092'4[B]   78-18096
ISBN 0 521 22297 4

For my daughter Monica
who made me write this

Contents

Preface

Why did I write this book? Very simple. On my seventieth birthday a couple of dozen of my old students, many with their wives, gave me a dinner party in Trinity College, Cambridge. Of course I had to make a speech, and I chose to give them a potted autobiography, compressed into twenty minutes and consisting largely of anecdotes. Afterwards various people suggested I should write those stories down, and my daughter was particularly insistent; so I got to work. At first I attempted to write down just the stories, but that didn't really make sense; there had to be some continuity. So I started with my parents and my childhood and told a few stories about my family; from there I continued through the rest of my career, including as many anecdotes as I could remember, in particular if they shed light on some of the great people I have met, such as Einstein, Stern and Bohr.

I don't have the total recall that makes the ideal autobiographer, nor have I kept diaries. I have always lived very much in the present, remembering only what seemed to be worth retelling. Even that ability is now partly gone; when I try to recall more recent memories they are fewer and less clear than those of my early days. This is not unusual, and it has persuaded me to stop my stories roughly at the time when I settled down in Cambridge, almost thirty years ago. Lots of interesting things happened during that time. I was right there when Perutz, Crick, Watson and others did that exciting work in elucidating the structure of protein molecules and the hereditary genes, the very stuff of immortal life. But I am ashamed that I remember little of that; apparently at the time I was chiefly irritated that those biologist chaps would occasionally borrow my instrument maker for a few days.

Also it is always a little tricky to write about things that are not very far back. Many of the people are still alive and may be offended about this and that, and although one can of course show them what one has written (this I have tried to do) it gets more complex as you come nearer the present. It was a very interesting time in the history of the Cavendish Laboratory, but much of it has been told by others and more will no doubt be told. So I don't feel too bad about giving merely a sketchy account of my life in Cambridge though I have lived here for thirty years.

While I never thought that my memory was in any way complete I used to think that it was reliable. But I have had some nasty shocks. Typical is the case of an orchestral performance which I heard at Cornell University in 1957 or 1958, and which I stoutly maintained had been given by the Cleveland Orchestra, famous for its virtuoso playing. But years later I found the programme of that concert, and it was a different orchestra altogether. I had simply come to believe that only the Cleveland Orchestra could have given such a marvellous performance. So I apologize to all those whom I have not consulted if they spot any inaccuracies, and I shall be glad to hear about them, even if only to get my own memory put right. Although I always try to relate a story exactly the same way each time it is still possible, particularly with something not told for a long time, to remember only the gist and to fill in the details wrongly.

I have always enjoyed making pencil sketches of people. They were not deliberate caricatures, but of course I tried to bring out the characteristic features – a large nose or a receding chin – and some exaggeration usually crept in. In the same way, when I write about some of the marvellous people I have met in my life, I am not attempting to caricature them when I tell just a few anecdotes, showing perhaps some odd quirk; it is just the same technique, as in a pencil sketch, of slightly exaggerating what I see as an amusing feature.

Don't regard this as a work of history. There are others who have been much better qualified to describe the events which the world has seen in my lifetime. I have always, as I already said, lived in the here and now, and seen little of the wider views. In this book I have chiefly tried to bring to life some of the people I have met, from what little I remember.

O.R.F.

Vienna 1904–1927

My paternal grandfather, Moriz Frisch, was a Polish Jew from Galicia who settled in Vienna and started a printing shop in 1877. He seems to have been one of the pioneers of the printed form. In the old days a lawyer had to employ a clerk (or several) to write documents in elegant copperplate lettering, starting with 'Whereas it be agreed between...' and so on and so forth. Moriz Frisch had the general text printed in copperplate writing so that the clerk only had to insert the names and addresses and other specific points of the contract. Naturally that brought him in touch with lawyers, and so it is not surprising that in 1902 his son, Justinian, married a daughter, Auguste, of a lawyer, Dr Philipp Meitner, who was also a keen chess player and a free-thinker with political interests. In his house my father met some of the people who later became well known in Austrian politics. I'll have a lot more to say about my father.

My mother was a pretty black-haired girl, the second of eight children; a precocious pianist who had played in public, or anyhow at one of the grand soirées given in wealthy Viennese houses, at the age of twelve. Alfred Grünfeld, Emil von Sauer and Theodor Leschetitzky were among her teachers. She also studied composition and conducting. There is a story that on one occasion, when a teacher repeatedly interrupted her conducting, she finally turned

1. The author with his mother, about 1910.

to him in mock desperation and declared 'Sir, if you interrupt me once more I'll throw myself into the orchestra!' and all the young men in the orchestra shouted 'Yes please!' In the course of her studies she wrote some quite ambitious music – a pretty good fugue, a fragment of a symphony, and things like that – but her best compositions were in a light Viennese or French vein; a younger brother – Fritz – wrote most of the lyrics. The manuscripts have largely been lost; as none was printed the music is just in my head, and nobody else plays it. Parts of an operetta survive; I think some of it is comparable to Franz Lehar or Leo Fall.

When my mother married she gave up piano playing for a while; she felt she couldn't practise when she had to look after me and entertain for my father who loved fun and had money to spend. (It didn't last.) Later she started practising again and on many occasions she would go to the piano (after a good deal of coaxing) and give us pleasure with her playing of Chopin and particularly Schumann whom she played extremely well. Much later, about 1931, she even attempted to resume her concert career; she played

2. Lise Meitner (1878–1968), aunt of the author. A pioneer in radioactivity, nicknamed by Albert Einstein 'the German Madame Curie'; born in Austria, but worked in Berlin from 1907 to 1938.

the Schumann Concerto on the Austrian radio, but had such stage-fright that she never tried again.

The next child her mother bore (in 1878) was Lise Meitner. (Not 'Lisa'. To pronounce it, say 'Leezet', but don't sound the t.) She was rather dominated by her two elder sisters, but also had to look after the five younger children that followed, and had no time for dress or things like that. Early on she became fascinated by physics and determined to study it, but her father made her first take a teachers' diploma in French, so that she could support herself if necessary. Only then was she allowed to prepare herself for the so-called Matura (similar to our A-levels), the examination that would qualify her to enter the university. She completed that training – which normally takes eight years in high school – in two years of very intensive work. Her brothers and sisters used to tease her: 'Lise, you are going to flunk, you have just walked through the room without studying.' But she didn't flunk; she was in fact one of the four girls that passed the examination, out of fourteen.

I remember very little from my own childhood. It seems that I was a bit of a prodigy who could speak, read and do arithmetic earlier than most other children. I could multiply fractions in my head when I was five, or so I was told; I certainly don't remember

3

it myself. In 1914, not quite ten yet, I was sent to high school, Gymnasium as it is called in Vienna; Latin was obligatory, and Greek for the first four years. I quite enjoyed Latin and remember reading some Tacitus just for fun; I thought he was a jolly good war correspondent, reporting on Julius Caesar's troubles in Germany. In Greek I nearly failed; but sometimes I was lucky. Once I was called in class to read out and translate a bit of Greek text; having read it and wondering what on earth it meant I was halted by the teacher who told the class 'Now boys, this is how Greek should be read. This is how one reads if one understands it; sit down, Frisch, no need to translate.'

But on the whole I have very little recollection of my school-days. The First World War meant poor food, some worry about my uncles and (worst of all) that I could not buy all the chemicals for my childish experiments; otherwise it passed me by. (My father was not deemed strong enough for active service.) I don't think I either liked school or particularly detested it, except that during the last year I was getting a bit impatient and wanted to be out and doing something. By the time I was twelve my mathematical gift had become fairly obvious, and from then on I was never called to the blackboard because my mathematics teacher knew I would pass every test he could think of; and in my last year but one I actually helped the boys in the form above to pass their final examinations (by fair means and foul).

During the more tedious lessons I used to play chess under the desk with Hans Blaskopf, the boy next to me. The men that had been moved were marked on a scrap of paper which soon became messy with erasures and crossings-out. Once as I was deliberating my next move and thoughtfully tapping the pencil stump I held between my teeth, it suddenly slipped into my mouth and down my gullet before I could stop it. Hans, unperturbable as always, merely offered me his pencil as a replacement. My mother was rather more agitated and took me to the doctor, who prescribed a diet of sauerkraut and mashed potatoes to ease the pencil along its tortuous journey, from which it eventually reappeared after several days, split lengthwise, the lead having been used up in blackening my inside.

I remember that incident mainly because it almost lost me the chance of attending a popular lecture by Albert Einstein; I had great difficulty in convincing my mother that the pencil was not yet due that evening. The lecture took place in a very large hall and I neither saw nor heard much of Einstein; but it was a great event for me. Soon afterwards I met a boy with similar interests and together we studied Einstein's popular exposition of his special relativity theory with great care. I have forgotten his name and don't know what became of him.

It was towards the end of my school-days, in 1922, that inflation hit Austria. It wasn't as bad as in Germany, where the value of the mark was halved about twice a week; but it was bad enough. I gave coaching for several months to a school-mate who had trouble with mathematics. I invented a system of mnemonics, little verses, which he had to learn by heart; with his good memory they helped him get through his examination better than we had hoped. His grateful parents doubled the fee which had been agreed in advance; I kept it in my pocket for a couple of weeks and then used it to buy a pencil, an ordinary wooden pencil; that's all I got for it.

I was never good at sports or games, although I was quite nimble and had a good sense of equilibrium; one of my favourite tricks was to jump on a running tram with both hands in my trouser pockets, and surprisingly it never led to an accident. I tried skiing but had not enough strength and courage; after twenty-five years of occasional skiing, mainly on practice slopes, I gave it up.

The only sport I did pursue was lawn tennis. At the end of my school-days I was one of the half-dozen or so committee members of a club which we had formed and which gave us a chance to play a reasonable amount of tennis without having to pay for it. What we did was to rent a court from one of the big clubs for the whole year; for that we got favourable terms. In order to recover our money we sub-let the court to other people, whom we recruited at a dance in the late winter for which we grandiosely rented one of the smaller ballrooms in the old Imperial Palace (die Hofburg) in Vienna. In the course of the dance every one of us tried of course to impress the young ladies with our skill as tennis players and with the delights of playing that game of kings; we offered our free services to teach anybody who didn't know the game or wasn't good at it. (I usually 'taught' every morning from seven to eight.) By the time the dance petered out in the small hours we usually had not only a list of new members but even the rudiments of a timetable.

There was one occasion when it nearly went wrong. I had dressed up as Charlie Chaplin, wearing a pair of very long trousers, a very short jacket and of course a bowler hat, which was too large and kept falling over my eyes. The role rather cramped my style as I didn't know how I should talk; Charlie Chaplin, in those days of silent films, didn't talk. At the end of the ball the list with all the names and rudimentary timetable could not be found. We all searched in our pockets, under the chairs, and in odd corners; we were pretty desperate to see all our efforts wasted. However, in the end we went home dejectedly; perhaps, with many phone calls, we could reconstruct the list in a week of hard work. At home when I took off my bowler hat a piece of paper fell out. It was the missing list; I had stuffed it into my hat to stop it falling over my eyes.

My father was a man of wit and knowledge, widely known in Vienna's intellectual circles as a stimulating companion and as a raconteur with a great store of Jewish and other jokes. He had a phenomenal memory and single-handedly produced, over a number of years, a sort of encyclopedia in instalments, the A–B Catechisms of General Knowledge. There were one hundred questions and answers in each slim issue of 16 small pages, with titles like 'Napoleon', 'The North Pole', 'Aviation', 'Gem Stones' and so on. There was just one I had to write (in 1935) entitled 'The Atom'.

3. Self-portrait of Justinian
Frisch (1879–1949), the
author's father.

His wit depended largely on puns or on amusingly distorted quotations; so it is difficult to translate. But I can translate one remark which was reported to me (for obvious reasons I was not present). When my parents were engaged they were of course never allowed to be alone together; there always had to be an older woman, a chaperon, to safeguard the virtue of the lady. On one occasion, however, the chaperon relented and allowed the two young people to go up one of the towers in Prague, pretending that she was too tired for the long climb. When the two came down again, flushed and very happy, she kindly asked 'Was it beautiful?' 'Oh yes,' said my father, 'we overlooked the whole of Prague'.

Of my father's childhood I know nothing. He went to the University of Vienna and got the doctorate of law. After that, for a while, he earned a pleasant living by travelling in Italy and painting hotels. Whenever he saw a promising one he would sit down and do an attractive water-colour sketch; soon a guest would appear and, after a while, the owner. When he duly admired the painting my father would gladly present it to him, and over subsequent drinks he would point out that the hotel might do much better if an attractive prospectus were sent out, extolling its beautiful location and excellent cuisine, and illustrating its charm with a reproduction of the painting. When the hotelier agreed bu

7

didn't know how to get such a thing done, my father would modestly say that he happened to know a printer who could undertake the job. The painting was sent home and the prospectus duly printed; my father got a rake-off that enabled him to travel to the next hotel.

His love for Italy soon persuaded him to try for a second doctorate, in the history of art. He seriously studied the Renaissance, and a good many books still on my shelves bear witness to his wide range of interest and in particular to his work on the origins of the palace of Urbino, on which his dissertation was centred. The dissertation was in fact written when fate struck; his own father went bankrupt.

Moriz Frisch had always been a generous man, proud of his great white beard, white at the early age of fifty, and never worried about money, of which there always seemed to be enough. He had given freely to anybody in need and was thunderstruck when creditors began to gather and he found he couldn't satisfy them.

To my father, who had to pick up the pieces, it was a dreadful lesson: never to be careless with money. For the rest of his life he would write down every penny he spent, and sometimes pass half the evening trying to trace a small discrepancy between his book and his purse. He also tried to make me do the same, but I never got into the habit, and I am still rather careless with money. Fortunately I have always earned enough for my modest needs.

My father never got his second doctorate and only just managed to save the firm, and that by accepting partners whom he didn't much like. I hardly remember them. I was only a small child and I vaguely remember the great thumping presses, the most modern of their kind, but very old-fashioned by present-day standards. The printing shop was a long tram ride from home, and I seldom went there. After a few years the disputes with his partners reached the breaking point; my father sold his share, but the firm kept the name Frisch & Co., somewhat to his later embarrassment when it became known that part of their business was pornography. Fortunately all my father's friends knew that he had broken with the firm long ago.

I have no recollection of the villa on the outskirts of Vienna where at first we lived; it had to be sold when I was still a toddler, and we moved into the centre of town, into a third-floor flat in one of the grimy blocks housing so many of Vienna's population. My narrow bedroom gave onto a light shaft; no grass or tree to be seen, and the nearest park (don't step on the grass!) a quarter of a mile away.

My father got an appointment at the biggest printing and publishing firm of Vienna (Waldheim-Eberle A. G.) and within a few years rose to be a director. But in 1924 the firm changed hands and my father had to leave. During the next twelve years he held a variety of jobs. In 1931 he started an advertising firm together with a young and promising draughtsman and a skilled photographer. It was a chancy business, and though he wrote some of the most quoted slogans to be seen on posters and newspapers the living was very modest. So he was glad when in 1936 he was invited to work as a technical expert for a publishing firm called Bermann-Fischer. The owner, Dr Bermann, was the son-in-law of S. Fischer, the head of one of Germany's great publishing houses, and had added Fischer to his name to stress the connection. He had some of the best German writers on his list – people like Thomas Mann, Franz Werfel, Carl Zuckmayer, Stefan Zweig and others. That was a good place to work. My father combined his experience of printing art, his thorough knowledge of type faces and his sensitive taste in helping Bermann-Fischer produce books which were widely respected for their distinguished appearance.

I owe a great deal to my father, and not only genetically. He had married young and so he was a good companion in my teens, on country walks and occasional excursions further afield. I was a sensitive child and I once burst into tears when he asked me why I was decapitating thistles; they had a right to live just as I had, he said. One warm spring evening I collected a dozen cockchafers, drowsy on the branches, into a twisted paper cone; my father didn't stop me. After coming home we watched those large brown beetles waking up and starting to move in the warmth and light of our room. They were locked in a hopeless tangle; whenever one of them was about to move away he was hooked back by another one. But finally one of them managed to disengage himself and slowly began to climb the smooth slope of the paper; and we watched with bated breath as he cautiously advanced, never moving one of his clawed feet before he had a secure hold with the five others. It took him several minutes to mount those two inches to the edge; then he rested a while before he opened his wings and buzzed out through the open window into the night. We felt we had watched a hero break away from the stultifying multitude, making his own way, against severe odds, to freedom. No sermon about human dignity and liberty could have done as much for me. I emptied the rest of the beetles on the window sill without another word.

My father had a strong feeling for right and wrong. He was not religious in the conventional sense, indeed he was very much against any kind of organized religion; but with his profound respect for life he might have been called a Buddhist and indeed took a great interest in Buddhist writings. I remember one occasion when he startled me by telling a visitor that I knew Pali, which of course I didn't. He then opened a copy of the speeches of Buddha in the original language, which the visitor had brought along, and showed me the first three words of a chapter, asking me what they meant. He had judged me rightly. I had previously glanced into a German translation of Buddha's speeches and knew that every chapter began with the same words; so I was confident in saying that 'Evam me sutam' meant 'This I have heard'. Fortunately the visitor did not probe my knowledge of Pali any further.

It was my father who aroused my interest in mathematics – real mathematics, not arithmetic. When I was ten years old he introduced me to Cartesian coordinates: he drew two straight lines at right angles on a piece of paper and told me how to attribute two numbers (coordinates) to any point by calling the distance from the vertical line x and the distance from the horizontal line y. Then any equation relating x and y would represent a whole set of points, in fact a curve. The next day I came back with the equation of the circle: $x^2 + y^2 = r^2$. (I admit it took me three weeks to figure out the equations of a straight line.)

When I was about twelve he introduced me to trigonometry. I still see his astonished face when, having defined sine and cosine, he wrote down (expecting to surprise me) the equation $\sin^2 x + \cos^2 x = 1$ and I said 'well, that's obvious'. This little story I tell you, not to boast but to point out my particular gift: speed. I don't think I have an original mind, but I used to be able to see quickly through logical connections and to go on to the next step, often ahead of others.

One of my uncles (Rudolf Allers, philosopher and member of the 'Vienna Circle') introduced me to Olga Neurath, a very remarkable woman. Years previously, a brilliant student of mathematics, she had developed terrible headaches with gradual loss of sight. Brain tumour was suspected; but after her eyesight had gone the headaches stopped; blind, she lived for many years. The young mathematicians in Vienna adopted her as a kind of communal godmother. They read to her and guided her when she wanted to go out; in turn she explained mathematics to them, and we all marvelled at her ability to do quite complex calculations in her head. She needed no help in lighting her pipe and always held her hand over the match afterwards to make sure she had safely blown it out.

I was about sixteen when I was introduced to her, and she taught me a great deal. At that time I had worked out the focal length of a concave mirror by an odd trick which I found exciting; indeed I had discovered the rudiments of differential calculus for myself. She showed me how such problems could be attacked in a systematic fashion, a year or more before calculus came up in school, and much more exciting through the way she guided my probing mind. She also introduced me to the ideas of four-dimensional geometry, and the idea fired me; in a few weeks I worked out the properties of all the regular polyhedra in four dimensions. The most complicated one, consisting of 120 penta-gondodecahedra, took several days of preparation and then two hours of unremitting concentration which gave me the first headache of my life. To visualize things in space does not come naturally to me – I am an auditory type and find it much easier to replay a piece of music in my mind than to visualize even a familiar face or scene – and those months of fascination with four-dimensional geometry were excellent training which later helped me, and still does, in designing complex scientific devices. I still think of Olga Neurath with affection and gratitude.

Her husband, Otto Neurath, was serving a prison sentence in Germany for having taken part in an abortive Communist rising just after the First World War. Later he was released and came back with all his books. A larger flat was found where the books were at first pushed into one room, filling it from the door to the top of the opposite wall. Then he worked for weeks with hammer and nails, covering all the walls with shelves, while we students took shifts in putting the books in order and standing them on the shelves, in two rows, one behind the other. They were mostly on sociology and socialism; I never understood the difference.

When I entered the University of Vienna in 1922 I realized there were a few other mathematical geniuses around; moreover, I began to feel that perhaps I wouldn't want to spend my whole life with a pencil and waste basket as my only tools. (Nobody even dreamt of computers.) I had always enjoyed making things; so I turned to physics and chose it as main subject, with mathematics second. In fact I read so little mathematics that I very nearly failed my final examination; what saved me was the fact that my examiner was rather deaf, and whenever I saw a puzzled expression appear on his face I carefully changed what I was saying to something more like the opposite. In the summer of 1926 I graduated to the D.Phil., a little before I was twenty-two. That was normal; it only took four years study in all (and only three years in economics!) to acquire a doctorate in Vienna.

After that I was a bit at a loose end. I had meant to go into industry; the art of making radio valves was developing rapidly, and the industry no doubt had need for young physicists. But I never got into that field. Instead I joined a very unorthodox little firm, run by Dr Siegmund Strauss, an Austrian inventor who had for a while worked with Telefunken and told me that he had made some very fundamental advances, such as feed-back and resistance-capacity coupling. He was full of ideas, and it was part of my job to listen and throw out the dud ones. He was very good-natured about it. Whenever I said why one of his ideas wouldn't work he would say 'Ah yes, you're quite right. Now let's see. Ah, we'll do it this way!' and off he was on a new idea. He had something like a hundred ideas every day of which only about one was any use; but a useful idea a day is really pretty good. The firm manufactured X-ray dosimeters (for measuring the intensity of X-rays) which went all over the world; X-ray therapy, demanding accurate dosage, was spreading rapidly.

For a while I worked in a conservatory, attached to his villa and separated from the elements by a thin glass partition; it was very cold in winter. My life there was an alternation between two evils. The gas fire gave me a roaring headache after a while because it had no flue and filled the room with carbon monoxide. When I couldn't bear the headache any more I turned the fire off. After another half hour the headache was gone but my toes were frozen; so I had to light the fire again. That went on for a number of weeks and doesn't seem to have done me any permanent harm.

After about a year, much to my surprise, I was offered a job in Berlin. That was quite a fluke. The job had been offered to someone who accepted but then withdrew when he suddenly found himself rich because a wealthy aunt had died. Then they offered the job to one who for some reason couldn't take it; finally, I don't know how, my name was suggested. Lise Meitner, who then lived in Berlin, was asked her opinion. 'Well,' she said, 'I can't really give you my opinion, he is my nephew and I may be biased.' They pressed her but she wouldn't give in. She insisted that her nephew must make his own way. In the end one of them asked her 'Tell me one thing. Is he a disagreeable person?' At this point she reflected a little and said 'No, a disagreeable person, that he is not.' So I got the job, probably on the recommendation of Professor Karl Przibram who had supervised my Ph.D. work; a kind and well-beloved man who many years later died in the fullness of wisdom, at ninety-three.

Atoms

Today we debate how many atomic power stations ought to be built; yet, incredible though it seems, at the turn of the century many respectable scientists did not believe in atoms. They admitted that the idea was useful to chemists in explaining their empirical facts, as John Dalton had suggested soon after 1800. For instance, pure marble contains exactly three times as much oxygen (as well as a fixed proportion of carbon) for a given calcium content as does quicklime; it was plausible that the smallest particle – one molecule – of marble contained three oxygen atoms for each of calcium while a molecule of quicklime contained only one. There were hundreds of observations like that; but from all their painstaking work the chemists would deduce only the *relative* weights of those hypothetical atoms, for instance that an atom of calcium weighs 2.51 times as much as one of oxygen. By the middle of the nineteenth century many of these (relative) atomic weights had been measured but nobody had any idea of what atoms really weighed; except that they were much too small to be seen or weighed individually.

In the mid-1860s several scientists tried by different ways to estimate the size of atoms, and came up with similar answers; from then on a small number of avant-garde physicists began to take atoms seriously. But what could one hope to learn about things that were only about one-hundredth-millionth of an inch in size? It seemed obvious that they could not affect the behaviour of matter in bulk; to the inventors and engineers who were the kings of the age they seemed irrelevant.

In addition I believe that there was a psychological block, based on a vested interest. Differential calculus, a new kind of mathematics which Newton and Leibniz independently invented, had been greatly developed and refined. That calculus allowed us to deal, not with things to be counted (like sheep or money, the job of arithmetic) or measured (geometry and algebra), but with things that gradually change; from the motion of planets as they continually change direction on going round the sun to the bending of a girder when it is loaded. Physicists rejoiced in their ever growing skill to solve rather artificial problems such as the vibration of an elastic metal plate of, say, rectangular or elliptic shape. If such a plate were an assembly of uncounted millions of atoms one might have to develop new, more complex and less elegant methods of mathematics. Why ask for trouble?

But there was worse trouble ahead. Not only our freedom to subdivide things indefinitely was to be curtailed if matter consisted of atoms; the freedom of those very atoms to move was to be curtailed. Otherwise it would require infinite energy to raise the temperature of any object even by a small amount! For some twenty years that absurdity and others like it were brushed aside even though they followed from the accepted laws of physics. Then, just at the turn of the century (December 1900) the worst difficulty, relating to the light emitted by glowing bodies, was cleared up by a conservative physicist aged forty, with a German Civil Service background, who was merely trying to put the theory of heat into an orderly shape. In doing that he started the greatest revolution since Galileo.

That revolutionary *malgré lui* was Max Planck. He calculated the rate at which a glowing body emits light of various colours; but he could do it only by assuming that light is emitted in energy parcels, which he called quanta. The energy content of one such quantum depended on the wavelength of the light, which for

4. Max Planck (1858–1947), the conservative German scientist whose 'quantum hypothesis' (1900) started a new age in physics. (Reproduced with kind permission of the *New York Times*.)

instance in red light is twice that of blue light, with the rest of the rainbow in between. To fit this theory with the observations he had to assume a new constant of nature; he called it *h*, and it became known as Planck's constant. For a light wave of given colour, the number of oscillations per second (the 'frequency') multiplied with *h* was assumed to give the energy content of each quantum. Without that assumption the calculation produced nonsense; with it he got accurate agreement with the existing measurements. So he published his 'quantum hypothesis', still hoping that it would soon be replaced by one that fitted into traditional physics. For several years nothing happened, except that more accurate measurements gave even better agreement with Planck's formula. So what was one to do? Planck, as one contemporary wrote, had 'explained' an incomprehensible fact by the even more incomprehensible assumption that light waves occurred in jerks.

It required an Albert Einstein to break that impasse, and he got the Nobel Prize for that, not for the relativity theory which he also published in the same year, 1905. He worked as a clerk at the Patent Office in Berne and created his theories in his spare time! Much of that paper is mathematical and of interest mainly to theoreticians. But it also showed that Planck's quanta accounted for some quite

5. Ernest Rutherford (1871–1937), son of a New Zealand farmer. In 1911 he discovered the atomic nucleus and split it in 1919. He was awarded the Nobel Prize in 1908, knighted in 1914, and created a peer in 1931. His ashes were placed in Westminster Abbey.

unrelated experiments by Philip Lenard which had remained inexplicable for several years. Here was more evidence in support of Planck's idea, and it gave quite independently the same value for *h*, Planck's constant. With two feet now firmly planted on the ground, Planck's brainchild could no longer be ignored.

A few papers were published which extended Einstein's reasoning, but it took another eight years before the floodgates were opened by a young Danish physicist, Niels Bohr, through his proposed model of the atom. You must have seen it many times, decorating almost any publication related to atoms: a dot surrounded by several circles, usually foreshortened into intersecting ellipses. That model has now been out of date for half a century. But symbols have long lives: Father Time is still depicted with a sand-glass, not a wristwatch.

Bohr's model was born in Manchester where he had gone to work with Ernest Rutherford, a New Zealand farmer's son who was awarded the Nobel Prize in 1908, raised to the peerage in 1931 and six years later buried in Westminster Abbey; a big man full of vigour, robust common sense and a passion to know. In 1911 he had deduced from experiments done in his laboratory that atoms were, in a sense, very much smaller than previously thought. More than 99.9% of the weight of an atom is concentrated in a minute central nucleus; the rest is empty space, patrolled by a few electrons. To Rutherford it seemed natural that those electrons would circle round the nucleus much as the planets circle round the sun.

That is where Bohr took over. He saw the need for some stabilizing principle which would guarantee that in atoms of the same kind the electrons occupied the same space; we would never get regular, well-formed crystals if their atoms came in all shapes and sizes. The key must lie, he saw, in Planck's quantum hypothesis, which would have to be extended so as to specify the electron orbits, just as in Planck's hands it had specified the energy content of light quanta. It took Bohr several months of hard work to decide how the extension must be done and what would follow from it. He tackled the hydrogen atom first, the lightest of all, containing only one electron. When his first paper on this appeared in July 1913 it accounted for so many observed facts that clearly there was some important truth in it; yet the details seemed absurd. Why should the electron be confined to specified circular orbits around the nucleus, as Bohr claimed? Was the hydrogen nucleus surrounded by a set of concentric rails on which the electron had to travel? And how did the elecron manage, on jumping from a bigger to a smaller circle, to push its surplus energy out as a light quantum? All this was totally alien to traditional physics.

And what about more complex atoms? Did the two electrons in a helium atom chase each other around the same circle, keeping the nucleus between them? Or did they describe concentric circles? Or did they travel on intersecting circles, with a timetable to avoid collisions? None of those proposals fitted all the facts. And why didn't electrons travel on ellipses as planets do? Did Planck's brainchild, now grown to the status of an atomic policeman, forbid it?

Gradually some order came into that chaos, largely under the influence of Arnold Sommerfeld, who taught at the University of Munich. Through his school passed most of the young theoreticians who later rose to prominence. Among them was a Viennese prodigy, Wolfgang Pauli; I met him later in life and shall tell you more about him. Sommerfeld said 'I can't teach him anything; at my suggestion he is writing a summary of Einstein's relativity theory'; it remained the best summary for many years to come. Pauli is widely remembered for his 'exclusion principle' which he proposed in 1923 and which earned him the nickname of Atomic Housing Officer. That principle does not allow more than two electrons to live on the same orbit. The possible orbits (now including certain elliptical ones) had been classified by Sommerfeld, and their orderly occupation according to Pauli's housing plan – as you considered atoms with more and more electrons – finally satisfied the demand for stability which had been the driving force behind Bohr's first atomic model, and made it clear why atoms – despite their minute nuclei – take up the space they do, and why they behave chemically as they do.

Much had been achieved in those turbulent ten years, and what I have written here is the merest thumbnail sketch of the highlights as I see them. I could fill a page merely by listing the physicists (and chemists!) who made significant contributions during that time to our understanding of atoms. Other people have filled volumes by trying to map the tortuous pathways – some mere detours, others blind alleys – by which a coherent method of quantum theory gradually emerged.

A method, not a picture. Today we no longer ask what really goes on in an atom; we ask what is likely to be observed – and with what likelihood – when we subject atoms to any specified influences such as light or heat, magnetic fields or electric currents. Those questions can be answered by the methods of modern quantum theory. But as to what really happens...to that question there is no reply. That at least is the widely accepted view towards which

Niels Bohr in particular has been guiding the development of quantum theory.

It was the German, Werner Heisenberg who, at the age of twenty-three, suggested that the theoreticians should give up trying to trace the movement of electrons inside atoms where they could not be observed; instead they should find mathematical correlations between observable facts. His teacher at Göttingen, Max Born, taught the young genius some of the mathematics he had been trying to re-invent, and himself made important contributions to the interpretation of the formulae. But it was Niels Bohr, with whom Heisenberg worked in Copenhagen for several years, who insisted on understanding what Heisenberg's mathematics really meant; it was under his influence that Heisenberg formulated and, with physical rather than merely mathematical arguments, justified the uncertainty principle that made his name a household word.

Physicists from all over the world came to Copenhagen to work with Niels Bohr and spread his way of thinking, and naturally I do the same, having worked there for five years. But I think it is quite wrong to regard the Copenhagen School as an establishment, set on perpetuating the views of its founder. Bohr's views have been debated and criticized for half a century, and no alternative has seriously shaken them. Newton may be accused of having suppressed the wave theory of light, simply by his great prestige and by not sharing the view of his older contemporary, the Dutchman Christian Huygens. Will the future level a similar accusation against Bohr? I doubt it; but only the future can tell.

Of course atomic physics was not created by the theoreticians alone; they need observations to support (or disprove!) their theories, and it is the experimenter's job to provide the observations. That distinction is comparatively new; in the old days a physicist usually had the mathematical and experimental skill both to formulate and to test his theories. Not always so: Michael Faraday – a Londoner without schooling who rose to world fame – was a superb experimenter with the vision to conceive the novel idea of electric and magnetic fields; but it took James Clerk Maxwell – Scotsman and first head of the Cavendish Laboratory, Cambridge – to create the mathematical theory of electromagnetism. Since then physicists have tended more and more to specialize either on techniques which allowed them to perform ever more difficult experiments, or on mathematical skills of increasing subtlety. But occasionally one gets a man like Enrico Fermi, the Italian genius who rose to fame in 1927 as a theoretician and then surprised us all by the breathtaking results of his experiments with neutrons and finally by engineering the first nuclear reactor. On December the second, 1942, he started the first self-sustaining nuclear chain reaction initiated by man and thus became the Prometheus of the Atomic Age.

But we must go back to the beginning of the century. At that time it was felt that atoms were talking to us, but in a code we couldn't decipher. Throw a little table salt (or some other sodium compound, like soda) into a gas flame, and it briefly flares yellow; a lithium compound makes it go pink, and so on. If you look at such a flame with a spectroscope (basically a glass prism with a slit in front) you don't see the usual rainbow-coloured spectrum but only a few differently coloured lines. Such lines are seen only when the atoms are 'excited' by heat (as in a flame) or by an electric current (as in a neon tube), and each kind of atom has its characteristic pattern of lines, its own 'line spectrum', like a finger print. Here was a simple method to detect the presence of a particular chemical element; and in the 1860s the scientists were very excited when they found many of our well-known elements in the sun and even in distant stars. Moreover, some physicists felt that those line spectra contained clues to the structure of atoms; but they were in a similar plight as the early Egyptologists, faced with thousands of clear-cut but unreadable characters.

In 1885 an unlikely character enters the scene: a Swiss mathematics teacher named Johann Jakob Balmer. He boasted that, given any four numbers, he could find a mathematical formula that

6. Werner Heisenberg (1901–1976) who, in his early twenties, laid the foundation to modern quantum mechanics and for many years continued to contribute to its growth. Nobel Prize 1932.

connected them, and a friend gave him the wavelengths of the four spectral lines that are characteristic of hydrogen atoms. Surprisingly Balmer came back with a very simple formula which fitted the measured wavelengths with uncanny precision. Balmer published his formula, but nobody could see any physical sense in it, and it remained a mere oddity.

But when Niels Bohr, in his struggle to adapt Planck's quantum idea to the structure of atoms, finally came across Balmer's formula 'everything became clear' as he later said. Within a month he had completed his great paper on the hydrogen atom, with the Balmer formula as its corner stone. Some formulae similar to Balmer's had been hesitatingly applied to the spectra of other atoms; they now made sense and could be used to test further extensions of the quantum theory. The cipher had been broken, and the atoms were revealing their structure.

Planck had merely assumed that light is emitted in quanta; Einstein had pointed out that it must also be absorbed in the same way, and a light quantum could not spread out in all directions but had to travel through space as a tiny parcel of electromagnetic energy, a photon as it soon came to be called. Bohr's idea of an atom as a tiny planetary system, with only certain orbits allowed, had been refined by Born and Heisenberg; one could no longer speak of definite orbits, but it was now possible to calculate the amounts of energy an atom was allowed to contain, and any change of energy (except in a collision) was accompanied by the emission or absorption of an appropriate photon. Within a decade the line spectra of atoms were well understood in most respects.

One of the last puzzles to be solved was that of 'fine structure'. With improved spectroscopes – under higher magnification as it were – many spectral lines turned out to be pairs of lines or even clusters of several lines very close together. Moreover those lines become again split up when the excited atoms are between the poles of a strong magnet. That effect of a magnetic field, found by the Dutchman Pieter Hendrik Zeeman in 1896, could be explained on the whole by traditional physics. Bohr's model, too, could explain it, but only by giving the quantum policeman yet another job: that of ensuring that the electron orbits formed particular angles with the lines of magnetic force. This 'space quantization' as it was called seemed to many physicists a very strange idea, and Otto Stern, then working with Max Born in Frankfurt, decided to test it by a novel method. I worked for three years with Stern in Hamburg, so you'll hear a lot more of him. His experiment – and many others – had become possible through the efforts of the lamp bulb manufacturers, of all people.

In ordinary air no molecule can travel more than a few millionths of an inch before bumping into another one. But in lamp bulbs the heat loss of the incandescent wire was minimized by pumping out most of the air. Even in the best 'vacuum' there are billions of molecules left, but pumps were developed which could reduce the number of air molecules in a container to one-millionth or less. The first proof that in such a vacuum atoms could indeed fly several inches without colliding was given in 1911 by the Frenchman Louis Dunoyer; but he did not pursue the matter. About 1920 Otto Stern decided to measure the effect of a magnet on a fine beam of silver atoms, obtained by stopping all the atoms evaporating from hot silver except those that passed through two fine holes in succession. When, together with his colleague Walther Gerlach, he carried out that test he found, as expected, that with no magnet the beam deposited a small dark spot of silver on a glass plate, placed in its way. When the beam was made to run along a strong magnet, shaped like a knife edge, that spot ought to have been broadened, according to classical physics; the silver atoms ought to have behaved like little magnets, oriented at random and thus randomly attracted or repelled by the knife-edge magnet. But in fact two spots were seen: the quantum policeman had done his job and ensured that half the silver atoms had become oriented so as to suffer full repulsion, the other half so as to get full attraction by the magnet, with none in between. Space quantization had been established; every student still learns about the Stern–Gerlach experiment. I was very lucky to get a job with Otto Stern in 1930, and the three years I worked with him were among the happiest and most fruitful of my life.

That experiment and the puzzles connected with the existence of the fine structure of spectral lines led to the last major step in the development of atomic theory: the discovery that the electron behaves, not just like a tiny ball of negative electricity, but as a ball that spins, and consequently as a magnet. That model accounted for the vagaries of the fine structure and how it was affected by electric and magnetic fields. It also explained an odd feature of Pauli's housing rule, namely that two electrons can share the same orbit: they can do so only if they spin in opposite directions!

At the same time (1925) an important new idea was suggested by a French student, Prince Louis de Broglie (pronounced 'de Broy'). Light, recognized as a wave phenomenon since 1800, had been shown by Planck and Einstein to consist of 'photons' which travelled through space like particles: perhaps electrons, recognized as particles since their discovery in 1897, might in some manner

7. Erwin Schrödinger (1887–1961). Austrian-born theoretical physicist whose wave equation helped greatly in understanding and calculating the behaviour of electrons in atoms. (Snapshot by the author.)

behave like waves? Experimental proof for this imaginative guess came within a few years; but even before that, the Austrian Erwin Schrödinger constructed his famous equation which was based on de Broglie's idea and gave the correct values for the energy levels of the hydrogen atom, just as Bohr's quantized orbits had done in 1913. Here was a new model for the atom, though not so easy to visualize; the electron now looked more like a pulsating cloud than an orbiting little planet. But for people who like to think in models it is still the best one we can offer. Mathematically, the Schrödinger equation turned out to be equivalent to Heisenberg's, though it looked quite different. Today we use one or the other, whichever is the best for attacking a given problem.

Schrödinger had great personal charm as well as mathematical gifts and really meant to be a philosopher; a number of his books, written with elegance, wit and clarity, show his preoccupation with fundamental problems. But the sensational success of his wave equation pushed him into physics. He was the only lecturer who ever made me feel like jumping out of my bench and yelling at him; he could present the arguments *against* some universally accepted

8. Paul Adrien Maurice Dirac. The Cambridge physicist whose mathematical skill and imagination set new goals for theoretical physics. His theory of the electron (1928) explained its spin, predicted the positron and earned him the 1933 Nobel Prize.

theory so vividly and convincingly that I felt sure the theory *must* be wrong! In hot weather he lectured open-shirted and wore tennis shoes at a time when professors in Berlin still stood on their dignity, by and large.

A year or two later, a young Englishman, Paul Dirac, improved Schrödinger's equation so as to conform with Einstein's relativity theory. It might have seemed a purely mathematical exercise, and it used some very unorthodox mathematics which Dirac himself invented. But it was splendidly successful: it accounted for the fine structure of spectral lines and the Zeeman effect; indeed it explained why an electron behaves like a spinning ball, with just the observed amount of spin and magnetic strength. It was a striking demonstration of the power of purely mathematical technique, and Dirac had set a pattern which has been followed ever since by confident mathematical physicists, with great success on the whole; though many an experimental physicist looks back with regret to the days when we still could follow what those mathematical chaps were up to.

But what about the atomic nuclei? All I have said so far relates to the outer part of atoms, to the space in which the electrons disport themselves, controlled by the electric attraction of that minute inner core at the centre of each atom, the nucleus in which over 99.9% of its mass is concentrated. Some facts regarding the size of nuclei and their vulnerability to fast bullets ('atom splitting') were obtained largely by Rutherford and his pupils, and other physicists gradually joined in the assault. But that is a separate story which I shall outline in another chapter.

Berlin 1927–1930

On my arrival in Berlin I had an odd experience; nothing very dramatic, and yet I had never been closer to death. When I got out of the station early in the morning there was little traffic as yet. I started to cross the road and jumped back as a cab honked at me. In Austria cars drove on the left (until Hitler's takeover in 1938). Berlin had right-hand traffic, and I would have to watch my step. Cautiously I crossed the road and then found that I had to get back again to catch a tram to the Kurfürstendamm. Stupid of me; this right-hand traffic kept fooling me.

The tram kept me waiting for a bit, and when it came I got on the front platform; there I was the only passenger and for a while just stood behind the driver, watching the trees of the Tiergarten fly past on both sides. There was a rap on the pane behind me; I turned round and found the conductor asking for my fare through a sliding window a few inches in size. He had trouble with my brand of German (and I with his) but after a bit he understood where I wanted to go and I got my ticket. Then I watched a long terrace on my left, tall blocks of flats, built in the pretentious style of around 1900.

The tram was travelling at a good pace, swinging from side to side; it was fun keeping my balance without holding on. I stepped back a little but did not actually lean on the gate on the side of the platform. On both sides of those trams they have gates, one

9. Otto Hahn (1879–1966), the German chemist and pupil of Rutherford, whose thirty years' joint work in Berlin with Lise Meitner led to many discoveries, including that of nuclear fission in 1938 for which he was awarded the Nobel Prize in 1945.

of which is removed to let people get on and off. The tram was swinging lustily now, and I was swinging with it, just avoiding touching the gate behind me. I could hear a lot of traffic at my back, businessmen in their big cars, passing our tram on their way to the Westend of Berlin.

Why had the gate on the left not been removed? Surely it hadn't been there when I got on? Or...how was that about right-hand traffic? In a cold sweat all of a sudden, I spun around, gripping the railing and staring at the cars overtaking us in a roaring stream, three abreast. All the time I had been standing and swinging with my back to a gate that wasn't there!

Lise Meitner helped me find lodgings; she lived in a tiny flat and couldn't put me up, but often asked me to dinner. Through her I met Otto Hahn with whom she had worked for twenty years, a collaboration that later became known to millions, leading as it did to the discovery of uranium fission, with the atomic bomb and atomic energy as the sequel. I always feel a glow of pleasure when I think of Hahn, with his bluff Rhineland accent and his self-deprecating sense of humour. He liked to whistle the last movement of Beethoven's violin concerto in an oddly syncopated manner; 'Doesn't it go like that?' he would say with assumed innocence when challenged.

I changed landladies several times but always stayed near where my aunt lived, in a pleasant suburb of Berlin called Dahlem. She worked (and later lived) at one of the Kaiser Wilhelm Institutes, the one for chemistry, and did radioactive research, of which I understood very little at the time. It was only much later that I realized what exciting things were going on there about 1927, just the time before the idea of the neutrino was first proposed, and at the time when quantum mechanics had just been launched on its conquest of atomic physics.

Every morning I had a long bus ride to the P.T.R. (Physikalisch Technische Reischsanstalt, equivalent to the National Physical Laboratory in the U.K., or the Bureau of Standards in the U.S.A.). It was a large building near the busy centre of Berlin, and a great many different kinds of work went on. I was in the optics division and my boss was Dr Carl Müller, not an uncommon name; a heavy, big man, slow spoken and quiet, but one of the most ingenious gadgeteers I have ever met. In his spare time he developed a process for making very thin metal films, supported, like drum skins, from the edge only; so thin that for instance you could read a newspaper through six of them, one behind the other! They were of gold but looked like ghostly coppery cobweb. He had some heart trouble and was laid up for a time; even in bed – his wife complained – he went on working with his soldering iron!

I had been hired to help him develop a new unit of brightness to replace the candle power, which was inaccurate and not very scientifically defined. The method had been proposed by Professor Emil Warburg who had since retired from being head of the P.T.R.; it was a very complicated method which I don't intend to describe.

In fact, Müller had a much more straightforward and more promising idea, but he couldn't work on it as long as his old boss was alive; that would have been a great insult. Soon after I quit, three years later, Warburg died and Müller started on his own idea. I don't know if it worked, but the 'candela' which is now the international unit of luminosity is not based on Müller's idea.

The head of the P.T.R. during the time I was there was a famous physicist, Friedrich Paschen, a very quiet and gentle person, whom I seldom saw. He had done very important work in atomic spectroscopy; a series of infra-red lines in the hydrogen spectrum is named after him. I heard only one entertaining story about him. It seems that one of his scientists came and asked for permission to put his beehives on the flat roof of the P.T.R. Paschen reflected for a bit and then said 'No, I can't allow that. It would set a precedent. Very soon somebody else would want to keep chickens on the roof, and after that someone would come with sheep; and before we knew where we are we would have cows on the roof of the P.T.R.' And that was the end of it; I don't know what happened to the bees.

Usually I stayed long hours trying out half-baked ideas of my own; then I took the bus home. Most of my ideas didn't work, but that's how one learns. We had a canteen, and I met a few of the other people who worked at the P.T.R. but I don't remember ever seeing the 'fever girls', which was the nickname of a dozen or so girls who worked all day calibrating thousands of clinical thermometers, one of the jobs of the P.T.R. I did meet someone who worked next door, by name of Walther Bothe (Nobel Prize 1954). At the time I didn't guess that he was doing very important work on radioactivity and in 1930 was going to find an important clue that later led to the discovery of the neutron early in 1932. The reason I met him at all was that he sent one of his technicians along to convey his request that I shouldn't whistle in the corridor because the whistling confused him when he was counting alpha particles. In those days we did not yet have automatic counting devices which would allow a physicist to read a meter at the beginning and again at the end of a run, minutes or hours later, when the meter would tell him exactly how many particles had traversed his instrument while he was doing something else. I often whistled Bach and the like, and it has occasionally got me in touch with chamber music enthusiasts; but on the whole my rendering of the Brandenburg Concertos was not popular.

Music has been important to me from childhood. My mother, as I mentioned, was a concert pianist and also a piano teacher, and she taught me from when I was about five years old. By the time I came to Berlin I played the piano with more panache than skill and loved to show off with Chopin scherzos and the like, really far too difficult for me. Lise Meitner added to my musical education, partly by making me play piano duets with her. She was not a good pianist and in fact I have never found anybody outside the family who knew that she played at all. But she had learned to read music, and together we slowly worked through things like Beethoven's Septet and other pieces which were melodious and where the slow movements were easy enough to give us real pleasure. The instruction 'Allegro ma non tanto' she translated as 'Fast, but not auntie'. (Aunt = Tante in German.) But she also introduced me to concert life, and through her I first heard the symphonies of Brahms and much of classical chamber music. It certainly opened my eyes to a wider world than I had known in Vienna where I heard my mother play but didn't go to concerts much; radio was still fairly primitive when I left Vienna in 1927.

As a child I disliked concerts, probably having been in a 'Konzertkaffee' where typically a three-piece band played popular numbers in a noisy, smoke-filled room to people who were chatting and drinking coffee. Then one day (I may have been fifteen) my mother persuaded me to go to St Stephen's Cathedral to hear something by a composer who had written finger exercises called 'Two-Part Inventions'. I expected some dull organ playing and singing; and I still remember the thrill of realizing – within a few

33

bars – that the St Matthew Passion by J. S. Bach was great music. It is one of the few moments of my childhood I truly remember; huddling in one of the huge fluted pillars, with that F-sharp in the second bar tearing at my heart. After that I began to go to concerts, but never took to opera.

While I was working at the P.T.R. my main contact with academic life was through the weekly colloquia at the university. We prided ourselves of having a front bench occupied mainly by Nobel Prize winners. There was Max Planck, tall and spare with the noble profile that later appeared on the German two-mark pieces and so became better known to millions than that of any other scientist. He never came to terms with the quantum theory he had begotten; its elements of uncertainty and non-causality went against the basic tenets of his orderly mind.

Einstein was there too, occasionally, but he travelled a good deal and was also seriously ill for a time. He too regarded the quantum theory as incomplete. I only met him once, in the entrance hall of the university when Lise Meitner suddenly stopped me with the words 'here is Einstein, let me introduce you'. I hastily transferred a pile of books from my right arm to my left and dragged off my glove while Einstein patiently waited with his hand held out, typically informal, seemingly quite relaxed. Yet he was under considerable strain, of which I was not aware. Idolized by millions who did not understand his theories, he was also under vicious attack from some of his colleagues, for precisely the same reason, and with an increasingly anti-semitic flavour. He had tempting invitations from abroad, but his friends, in particular Max Planck and Walther Nernst who had persuaded him to come to Berlin in 1914, passionately pleaded with him not to leave them, and so did many others who fully recognized that he was the greatest physicist in Germany. He finally left late in 1932, just before Hitler came to power.

In his later years he usually wore a turtleneck pullover. In fact he detested all formality, and for that reason – so I was told – did

10. Albert Einstein (1879–1955). The first to take Planck's quantum hypothesis seriously. His paper of 1905 gave evidence for the existence of light quanta (photons), and it was for that paper – not the one in the same year that created the relativity theory – that he was awarded the Nobel Prize in 1922.

not stay in England, when Hitler came to power, but went to America instead. To give him a big welcome to their country, his English friends invited him to parties where everybody wore tail coats or dinner jackets, and food was served by liveried servants. Einstein felt he couldn't possibly live in a country where so many formalities had to be observed; so all that hospitality was completely self-defeating.

Einstein had wonderful powers of concentration, and I am convinced that there was his real secret: he could think for hours with the kind of complete concentration of which most of us are capable for only a few seconds at a time. Once someone he had made a date with in Prague had forgotten and arrived two hours late. He came rushing up to Einstein who was sitting on the parapet of the bridge where they had agreed to meet. His apologies were waved aside with the comment 'It's quite all right, don't worry; I can think here just as well as anywhere else.'

The following story was told to me by Gabriele Rabel, a writer,

35

philosopher and teacher who died in cheerful poverty (like Diogenes) around 1964 at the age of eighty, not far from Cambridge. In her youth she had enough money to travel all over Germany, studying at various universities. According to her account, once in her student days she addressed Einstein with admiration 'What a wonderfully quick mind you have, Professor Einstein! When those bright young men in the seminar report some brand-new theories, how is it that you can immediately put your finger on the sore spot?' Einstein replied 'My dear young lady, I cheat. Because, you see, those theories which the brilliant young men propose, I know them all; I have thought of them myself. So I know exactly where their sore spots are!'

Another story (I don't know who told me this one) relates that Einstein once gave a talk on some new idea of his at a meeting of the German Physical Society in a town now in East Germany. After he had spoken and the chairman had respectfully invited questions a young man got up at the back of the hall and started in bad German and a most surprising manner. He said something like 'What Professor Einstein has told us is not so stupid. But the second equation does not strictly follow from the first. It needs an assumption, which is not proved, and moreover it is not invariant the way it should be...' Everyone in the hall had turned round and was staring at the bold young man. Except Einstein; he was facing the blackboard and thinking. After a minute he turned round and said 'What the young man in the back has said is perfectly correct; you can forget everything I have told you today!' By the way, the young man was Lev Davidovich Landau, who later became the foremost theoretical physicist of the Soviet Union.

There are many examples of Einstein's endearing modesty and his willingness to admit errors. On one occasion, when a colleague reminded him that he had said something quite different three week previously, Einstein retorted 'Do you think God cares what I said three weeks ago?' In one of his letters to Max Born he admits having made a blunder in one of his calculations and adds 'Gegen das Böckeschiessen hilft nur der Tod' ('Only death helps against blundering').

Nernst was another of our front-benchers. Often after a lecture he would get to his feet – which made little difference as he was rather short – and, waving his hands, would say with his cracked old voice 'but that is just what I said forty years ago!' The students thought that funny, but he probably had. One of the great principles of physics, the third law of thermodynamics, was his creation, and in the early part of the century he had great influence with the Kaiser; indeed that famous chain of scientific laboratories known as the Kaiser Wilhelm Gesellschaft had been his idea. His invention of a new type of electric lamp he sold for a fortune before it was superseded by the tungsten filament lamp. His usual air of naive innocence was deceptive; he bought real estate and remained well off to the end of his life, despite the run-away inflation of 1922 when the value of the mark was halved twice a week for several months.

But Nernst made massive and lasting contributions to technology as well as science; modern batteries and fuel cells rely on his and his school's work on electrochemistry. For all the malicious stories that survive him we must not forget that he was one of the leaders of science in the first half of this century.

Gustav Hertz, a great-nephew of the famous Heinrich Hertz who had discovered electromagnetic waves in 1887, was quite bald; his head, shining like a polished billiard ball, was said to illuminate the darkened lecture room and make projection difficult. He had a peculiarly impish sort of humour. Sometimes he came to drink tea with the chemists in the laboratory where Hahn and Lise Meitner worked. On one occasion he waved the tea aside with the remark 'I am fed up with that stuff, give me the alcohol' and got one of the students to hand him a bottle of absolute alcohol from the shelf. Lise Meitner was horrified 'But Hertz, you can't drink that, it's pure poison!' Hertz took no notice, poured himself a

37

11. Gustav Hertz (1887–1975) who, with James Franck, won the 1925 Nobel Prize for experiments that showed in 1912 the existence of quantized energy states of atoms; great-nephew of Heinrich Hertz, discoverer of electromagnetic waves in 1888.

tumbler and drank it down without any ill effects; he had got the student to fill the bottle with water beforehand.

My employment at the P.T.R. lasted for three years, but during the last year I worked part time at the physics department of the University of Berlin in the section of Professor Peter Pringsheim. This is how it came about. According to the quantum theory an atom usually has some rotation (like a spinning top), partly due to its electrons running in circles and partly due to their spinning on their own axes. When an atom sends out a light quantum this rotation (this angular momentum) changes, and I had thought of a way in which I could cause the light quantum to take that extra momentum away by rotating like a cart wheel, about an axis at right angles to its travel. There was no room for such a motion of the light quantum in the theory, and so I was rather excited and wanted to try the experiment which, if the outcome were positive, would revolutionize physics. Lise Meitner got me in touch with Pringsheim (whom she knew well), and he gave me the facilities of his laboratory and permission to work there in the evening.

The first thing I had to do was to get a skeleton key; it appeared that all keys in the physics department had been lost long ago, and every student used a skeleton key to get into his own room, and into others when needed. I still remember the suspicious look I got when I tried to buy a skeleton key in a hardware store! But then one of the other students showed me what a skeleton key looked like, and I managed to forge one out of strong steel wire; it worked very well. I never joined the competition among the students as to who could open most quickly the most difficult doors, but I certainly learned to open my own room in a few seconds, except once when I came back after a holiday period and it took me five minutes to open the lock; somebody had inserted a gadget to make it burglar-proof. (Perhaps I have missed my true vocation.)

The experiment gave a perfectly clear negative result, and when I discussed it later with James Franck he told me where my reasoning had gone wrong. I was a little ashamed not to have spotted this (but neither had a theoretical physicist with whom I had discussed my experiment beforehand). My paper was already written, and the best I could do was to add Franck's explanation as a footnote. All the same, the whole thing had got me in touch with Pringsheim, and he gave me employment when my grant at the P.T.R. ran out. The work I did for him was at first of a very practical kind. The physics department had been in use for years when mercury (quicksilver) was used a lot, and pounds of it had been spilt and were hidden in cracks under the floor. It had just been realized that mercury vapour was highly poisonous; clearly there was need for an instrument that could rapidly test the air of a room for its content of mercury vapour. Pringsheim had worked out how that instrument should be made – it depended on the absorption of ultra-violet light by mercury vapour – and I built what he wanted. It worked, but unfortunately was also sensitive to vapours of harmless chemicals such as benzene; so it was not an unqualified success. But I learned a lot of technique, and later Pringsheim gave me a task which was of some scientific interest and which, since we had worked closely together on it, we then published together.

Our joint work was done in Pringsheim's room, and he had bare copper wires carrying electricity from one outlet to the other. When I pointed out that this was dangerous he shrugged his shoulders and said 'You'll just have to watch out.' As far as I know nobody ever got electrocuted there, although it might easily have happened if someone had been careless. In fact I believe fatal accidents are less likely in a laboratory where people know that they have to be careful than when they think that all is safe; it never is!

Hamburg 1930–1933

In 1930 I was offered a job in Hamburg – a real job for the first time, not just a grant. I think Pringsheim recommended me highly; at least I learned that a few months later he recommended a student of his very warmly, but added 'though he is not another Frisch'. Lise Meitner was so pleased to learn of that flattering clause that she told me, at the risk of making me even more conceited. One foot at last on the academic ladder! My boss was to be Professor Otto Stern, famous through the classical Stern–Gerlach experiment, and I was thrilled at the chance of working under that great physicist. I was to be what is known in Germany as *Assistent*, a sort of high-class technician. I never gave any lectures in Hamburg; for that I needed the '*venia legendi*'. Stern was about to ask the university to offer me that permission to teach when we both had to leave because of Hitler. My duties were to assist Stern in taking measurements and later in designing some of the apparatus which we used in our researches.

My first recollection of the laboratory is standing in front of what looked like a forest of glass, a sort of glass-blowers' nightmare; tubes and bulbs and cylinders and mercury pumps blown from glass, with stopcocks by the dozen connected in a manner that made no more sense to me than the twigs in a hedge. And there I watched Stern and his chief assistant, Immanuel Estermann, turning stopcocks apparently at random, closing this one and then after a few seconds opening that one, and so on for what seemed like half an hour. I felt I would no more learn this than a totally unmusical person would ever learn to play the organ. But actually it was only a few weeks before the thing made sense and it was fairly obvious which stopcocks had to be turned and in which order.

Estermann assured me that mistakes did of course occur and that there were in fact two basic 'experiments' which no physicist could perform too often: one was trying to pump air through a stopcock that was shut, and the other was trying to create a vacuum when one of the stopcocks to the outer atmosphere had been left open. Actually we did not make those mistakes very often.

A few months later Estermann went on sabbatical leave, and I moved into his place as Stern's spare pair of hands. Stern was rather clumsy, and moreover one of his hands invariably held a cigar (except when it was in his mouth); so he was disinclined to handle any breakable equipment and always left that to his assistants. I still remember what he did when anything appeared to topple. He would never try to catch it; he lifted both hands in a gesture of surrender and waited. As he explained to me 'You do less damage if you let the thing fall than if you try to catch it.'

Yet Stern was, in a higher sense, a superb experimenter. In using a new apparatus he left nothing to chance. Everything had been worked out beforehand and every detail of the performance was carefully checked. Stern would calculate, for instance, how much beam intensity he expected to get, even though that was a very lengthy and tedious calculation, which he always did himself. He could not predict the intensity very accurately; but if it fell short by more than 30% he felt that something must be wrong, and the fault had to be tracked down. I have never seen anybody keeping such strict control of his instruments, and it surely paid off. As a rule the experiments we did were so difficult that nobody else in the world was attempting them. That created an oddly relaxed atmosphere. Once, having found a very striking phenomenon which we didn't understand, we left it lying for a year in the hope

12. Otto Stern (1888–1969) the German physicist who made a versatile tool of a French invention: he used beams of atoms and molecules, first produced by Louis Dunoyer, to study many basic problems in atomic physics.

that we might yet think of an explanation before we decided to publish it. (A couple of years later an English theoretician explained it.) The idea that anybody could do the same experiment and beat us to that important result didn't enter our minds, and nobody did.

Usually our experiments required a number of controls to be handled at the same time, and it was essential that both of us should be sitting at the controls and cooperate closely. On one occasion a visitor commented on the fact that so many controls were needed, and Stern answered 'Yes, it is a pity that we lost our prehensile tails some millions of years ago, they would come in useful now.' Whereupon the visitor answered 'It wouldn't help you, Stern; you would just make your apparatus still more complicated!'

I must tell you a little about what we were doing. In any gas (such as air) the molecules keep bouncing around; a molecule in air will travel in a straight line for only a minute fraction of a millimetre before it collides with another one. At atmospheric pressure air is far too crowded with molecules. But pumps for

43

removing all but one-millionth of the air from a container had been known for about a hundred years; in such a crude 'vacuum' an air molecule could travel something like half an inch before making its next collision. At the turn of the century – largely because a better vacuum was needed in the early electric bulbs – vacuum pumps were further improved and it became possible to give gas molecules a free path of many inches. In 1911 a French physicist, Louis Dunoyer, was able to show that you could get a narrow beam of molecules (he used sodium atoms) flying through an evacuated container, each atom travelling in a straight line for several inches, the length of his glass tube. But he didn't realize that those 'molecular beams' were a powerful research tool, for studying various properties of single molecules.

It was Stern, employed as a theoretical physicist in Frankfurt, who started experiments to measure first the mean speed of gas molecules and later their deflection by suitably shaped magnets – designed by his colleague Walther Gerlach – in order to learn about their magnetic properties. That was the work – the Stern–Gerlach experiment – that made them both famous, by confirming some rather incredible conclusions from the quantum theory, still quite new then. He realized that there were many other phenomena to be studied in that way, and when he got his chair as Professor of Physical Chemistry at the University of Hamburg he set to work. The researches of his school were published in a series of papers entitled U.Z.M. 1 to U.Z.M. 30, using an abbreviation for a German phrase (Untersuchungen zur Molekularstrahlmethode) which means research on molecular beams. In 'U.Z.M. No. 1' (1926) he outlined a programme of research which indeed he largely carried out over the next seven years, until Hitler came to power. His amazing foresight enabled him to map out practically all the experiments that were feasible and important. He is still regarded as the great-grandfather of molecular beams; by now I have become one of the grandfathers, together with a few other pupils who have in their way spread the use of that versatile research technique. It is still almost a family affair, and nearly everybody who works with molecular beams can trace his 'ancestry' back to Stern.

Let me describe one of Stern's experiments in which I took part. As early as 1927, de Broglie's idea that a beam of electrons behaves like a train of waves had been confirmed by having such a beam strike the surface of a crystal and showing that the electrons came away in certain definite directions; the wavelength of the electrons calculated from this 'diffraction' fitted de Broglie's formula. But would a beam of atoms or molecules – each consisting of several parts – also be diffracted? Stern and his co-worker, Immanuel Estermann, had found strong indications that it would; but Stern wanted to make a real precision job of this and asked me to join him when Estermann went on sabbatical leave.

The helium atoms or hydrogen molecules we used came out of a nozzle into our vacuum with speeds varying around a mile a second; to select atoms of a particular speed we copied the trick by which the Frenchman Armand Fizeau had measured the speed of light in 1849. We 'chopped' our beam by making it pass through the edge of a spinning metal disk where out instrument maker had laboriously cut 400 radial slots. As each slot passed across the beam it let some atoms through; after a few inches they met another disk mounted on the same axle, and only those of the desired speed would arrive just in time to find a slot again to let them pass. They then hit a crystal; some were reflected as from a mirror, others diffracted, and we could measure the angles with a delicate movable beam detector (which I would love to describe but won't).

When Stern computed the wavelength of our beam from our measurements it differed from de Broglie's formula by only 3%. Most physicists would have rejoiced; but not Stern! He had designed our apparatus to an accuracy of 1%; something was wrong, he said. We spend days remeasuring all the dimensions; all correct. In the end, as a forlorn hope, Stern asked me to check that there were really 400 slots, and after hours of counting and recounting the incredible was clear: the precision lathe on which the slots had been cut was at fault, and there were 408! That wiped out 2% of our discrepancy, and the remaining 1% Stern was ready to accept.

We all used to go to lunch together, Stern and his four assistants and several others from the physics department. Stern was well-to-do and liked to eat well; so did we, but it had to be cheap. For a while we would eat cheaply until Stern convinced us that the food wasn't worth eating, and that he had found another place which was much better and only slightly more expensive. Then again, after some time, one of the young men said that he simply couldn't afford to go there and had found a restaurant where the food was almost as good and much cheaper. Stern was very good-natured about all that. Altogether his good humour and friendliness were his most pronounced characteristics. He was medium-sized and tubby, with black curly hair, getting thin on top, a chunky nose and a rather long chin; not an attractive face in itself but made attractive by his sunny smile and the intelligence beaming from his eyes.

The conversation at lunch was either about physics or about cinema. Stern went to the movies practically every night and sometimes saw two different shows on one day. He used to complain that none of the Hamburg newspapers would employ him as a film critic. He felt he could make a much better job of it, and he wouldn't charge anything because he didn't need the money, and he was going to the movies anyhow! So whenever we asked which film we ought to see, Stern would settle back in his chair with a happy smile and proceed to lay down the law on what was good and what was bad in the movies on show. When Hitler came to power Stern left Germany and settled in the U.S.A.; eventually he retired to Berkeley, California, where he lived to a ripe old age of over eighty having once evaded death by a heart operation. When death finally found him, very appropriately it was in a cinema.

Stern didn't like working late at night; six o'clock was usually

13. Wolfgang Pauli (1900–1958) the Austrian-born theoretician whose 'exclusion principle' established order among electrons and other particles. The existence of the neutrino which he proposed in 1930 was ultimately clinched in 1956.

the time we finished. But occasionally, when measurements were going well, Stern would go on, and if it got later than seven it was understood that he would take me out to dinner. He usually took me to one of the top restaurants in Hamburg, in particular one called the Halali (pronounced Hullulleeh, a huntsman's call). On one of those occasions I remember sharing the dinner with the legendary Wolfgang Pauli, on one of his occasional visits from Zurich. Pauli was a plump, youngish man, only a couple of years older than I, but had created a sensation when, still as a student, he had written the best article (for many years to come) on Einstein's relativity theory, and he had done a great deal to lay the foundations of quantum mechanics. One odd characteristic was that he kept rocking forward and backward, not only when he was sitting but even when he was walking. Since this rocking motion didn't keep time with his legs his walk was erratic; for a few steps

47

he would walk very fast and then the oscillation would interfere with his leg motion and a few short steps would follow.

Pauli's rudeness to his collaborators was taken for granted as part of his personality; none of them was put out on being called an incompetent idiot in public if Pauli had found an error in his work. Even Niels Bohr, usually treated with a modicum of politeness, was not safe from Pauli's sarcasm, as in this letter to Bohr's wife '... you wrote two weeks ago that Niels would answer my letter on Thursday; but you didn't say which Thursday. However, a reply written on any other day of the week would be equally welcome...'

If teasing is a mark of friendship then indeed Stern and Pauli were extremely good friends. Pauli teased Stern without mercy. I remember the first time Stern took both Pauli and me to the Halali. Pauli told Stern that he did this only out of the bad conscience of a rich man, a bad conscience which he tried to appease by spending insignificant sums of money on his less affluent friends. He was articulate in a biting sort of way and could run rings around Stern, who only chuckled in his good-natured manner and didn't really try to defend himself.

There was an odd legend attached to Pauli. The so-called Pauli effect was a kind of evil eye: it was alleged that when Pauli appeared anywhere near a laboratory a dreadful thing was likely to happen. Bits of apparatus fell to pieces or exploded, and so on. One story asserted that James Franck, working in Göttingen, came to the laboratory one morning and found that the cooling water had failed, that the pump had blown up, that there was glass all over the floor; an absolutely terrible mess. Franck's instant reaction was to send a telegram 'PAULI, WHERE WERE YOU LAST NIGHT?' The answer came back 'TRAVELLING FROM ZURICH TO BERLIN'. (The train passed through Göttingen.) Of course, I don't believe a word of that story, but it is typical of the legend. Once, in Hamburg, Pauli was invited to visit the observatory, an invitation which he at first declined with the words 'No, no; telescopes are expensive.' The astronomers smiled and assured him that the Pauli effect had no power in the observatory. When Pauli entered the dome there was an ear-splitting clatter; when the party recovered they found that the large cast-iron lid had fallen off one of the telescopes and shattered on the concrete floor.

One more story. I was working in my laboratory with our glass-blower, sitting on the floor and holding up part of the glass-work which he was going to weld to the rest of the instrument with his blow torch; a very delicate operation. At that moment the door opened and Pauli looked in. Nothing happened; I kept my nerve.

Hamburg is the one place where my habit of whistling did pay off. I was whistling one of my favourite Brandenburg Concertos in the chemistry laboratory when one of the chemists turned round and said 'Do you play the violin?' I didn't play well, but I said 'Yes'. He asked me along to play with him and some of his friends, and from this grew a long association with several people with whom I played chamber music about once a week, either the violin or the piano. In this way I got friendly with a most interesting person, the architect Otto Strohmeyer. He probably had the greatest variety of gifts that I ever found combined in one person. He was a successful architect, he had produced some impressive sculpture, did lightning drawings and silhouettes (in fact he told me that for a while he made his living by doing silhouette portraits with black paper and a pair of scissors, going from one pub to the next), and in addition he played several musical instruments. With us he played the cello but I know that he also played the organ and the bassoon. His competence in mathematics and astronomy was considerable, and he also gave very good popular lectures on those subjects; and he wrote very skilful light verse. He came from the south of Germany, from Bavaria, and spoke with a Bavarian accent which I found very refreshing, as it reminded me of my native Austria; the Hamburg accent seemed stilted in comparison. He had remarkable success with women even though he was small of stature and limped because one of his legs was short.

One of his stories is worth retelling here. He had designed a large, self-supporting roof for a customer, but the building inspector had rejected the plan; a roof that large must have pillars, he said, waving aside Strohmeyer's static calculations. So the plans were changed, and when the completed building was inspected, there were the pillars. But after the inspector had gone they were removed, crumpled up, and thrown in the dustbin: they had been cardboard dummies.

Sometimes, when we were in the mood, after playing chamber music until midnight, we would all pack into his small car and drive out to the Reeperbahn (it means 'ropers' track'), the red-light district of Hamburg. Not with any immoral intention but because Strohmeyer knew so many interesting characters such as barkeepers and performers of all kinds, having worked the district in the days when he made a living by doing silhouette portraits. We would go from one place to another, chat to the *habituées*, to the girls, to the barkeeper, sometimes getting free drinks, until it was four in the morning. At that time all the bars had to shut, but fortunately there were other localities which opened at that time and served delicious chicken soup, thought for some reason to be appropriate for sobering up. Finally Strohmeyer would drive us all home, one after another.

There was one occasion when I had confessed shortly before midnight that I was about to reach the age of twenty-eight; so we played Mozart's string quartet K. 575, a particularly festive piece which we always played on birthdays, and decided to celebrate in the usual manner on the Reeperbahn. Strohmeyer then told me that he had to go off to the Danish frontier in the morning to fetch his wife and his sister-in-law; would I come with him? So after bringing everybody else home he took me to his flat; we had some black coffee and started out at seven in the morning. One of my friend's collaborators, a young architect, was picked up, and the two of them took turns in driving, alternating every fifty kilometres. Strohmeyer slept soundly whenever he didn't drive. It took us three or four hours to get to where his wife and sister-in-law were waiting with three large trunks and an even larger dog. They were all crammed into the car, and we turned towards home, driving down this time on the Baltic side of the Jutland peninsula. Somewhere near Kiel the car developed a flat, and the journey became a nightmare. The spare tyre was no good; every half hour we had to stop and use the hand pump. It was eleven at night when we came back to Hamburg, and close on midnight when I finally dropped into bed. But I was on the tennis court at seven in the morning as I had promised my partner.

I have never been politically conscious. In Vienna I joined a political student organization for a short while, but I merely served on the entertainment committee, helping to arrange dances, and on the few occasions when I took part in political discussions I merely thought they were rather ridiculous. Brothers addressing each other solemnly as Mr Huber; meetings to receive delegates from other organizations, and so forth; I didn't take it seriously. Once we stood around for an hour, two dozen of us, waiting for a German delegate whose train was late. When he came he went to each of us in turn, clicked his heels, stuck out his hand for a shake and said 'Bock' (his name). It had a hypnotic effect on me, and when he came to me I responded by also saying 'Bock' instead of 'Frisch'. He made no comment about us being namesakes; probably thought it undignified to do so.

In the early thirties in Hamburg I didn't pay any attention to the general crisis atmosphere; with a sarcastic smile I observed the repeated changes of government and the much joked-about ineptness of Hindenburg, the famous general who had been made President of the Republic of Germany. When a fellow by name of Adolf Hitler was making speeches and starting a Party I paid no attention. Even when he become elected Chancellor I merely shrugged my shoulders and thought, nothing gets eaten as hot as it is cooked, and he won't be any worse than his predecessors.

There of course I was wrong. It soon became clear that his

51

anti-semitism was not just talk, and when his racial laws were passed, Stern was quite shocked to find that I was of Jewish origin, just as he was himself and another two of his four collaborators. He would have to leave and the three of us as well, only one of his outfit – Friedrich Knauer – being Aryan and able to remain in a university post. Actually the university in Hamburg – with the traditions of a Free Hansa city – was very reluctant to put the racial laws into effect, and I wasn't sacked until several months after the other universities had toed the line. At first I still hoped that I might be able to take up a Fellowship which the Rockefeller Foundation had awarded to me (at Stern's instigation) to enable me to work in Rome for a year, an opportunity I had very much looked forward to. In Rome, Enrico Fermi had started a centre of scientific research which was rapidly gaining fame on account of his own genius and of the brilliant people he had started assembling around him. But the Fellowship was conditional on my having a permanent post to go back to; when Hitler's laws came into effect the Rockefeller Foundation regretfully informed me that under those circumstances they could not offer me the Fellowship any longer. Something else had to be found, and I remember how Stern went off to look for posts outside Germany for his collaborators. He was in no great trouble himself; he had private wealth, he was famous and would have no difficulty at all in getting a job, and indeed he joined the Carnegie Institute in Pittsburg soon afterwards and took Estermann with him. My other non-Aryan colleague, Robert Schnurmann, went to England; when I last met him he was a lecturer at Birmingham University.

In that fateful summer of 1933 Stern went to Paris and said he would try to get me a job at the Institut du Radium, where Marie Curie was the reigning Queen. When he came back some weeks later he told me that Madame Curie had no place for me, but that he had persuaded Patrick Blackett in London to offer me a place, and that the newly founded Academic Assistance Council (later re-named the Society for the Protection of Science and Learning) was going to give me a grant of £250 for one year, quite adequate in those days.

Those few months, from when I knew I had to leave until I actually left, were an odd period. In spite of all the uncertainties I managed to complete a piece of research which was published as U.Z.M. No. 30, the very last one in the series; I had measured the speed (about 1 inch/second) with which a sodium atom recoils on emitting a quantum of its characteristic yellow light, and that was a very direct proof of the particle-like behaviour of light quanta. Disturbing rumours were rife. Some of my Jewish friends had warned me not to be out at night because Jews had been beaten up in the dark. I remember walking home late one night when I heard fast footsteps ring out in the empty street; I wondered if it was one of those anti-semitic brutes on the rampage. Of course to break into a run would have given me away at once; I kept my speed though the footsteps rapidly came nearer and finally pulled up beside me. A burly fellow in S.A. uniform pulled off his cap and greeted me with great politeness; it was the son of my landlady. He explained to me that he had to join this para-military force because otherwise he would not have been allowed to complete his law studies; there were many like him who disliked the Nazis but couldn't afford not to join.

The persistent stories of concentration camps, of synagogues burnt, of beatings and torture, all were stoutly denied by the German newspapers as mere 'horror propaganda' put out by the enemies of Germany. Some of my friends told me the stories were true, indeed that the truth was worse. But I wouldn't believe that Germany had changed so suddenly and so horribly, and that all the newspapers could so consistently be telling lies.

In the summer of 1933 Niels Bohr invited me to his usual summer conference in Copenhagen which this time he proposed to use as a sort of labour exchange which might help those physicists who had to leave Germany in finding jobs abroad. I remember travelling north in the train from Berlin, opposite to a

14. Homi J. Bhabha (1909–1966) of a Parsee family, he became director of the Tata Institute in Bombay and finally chairman of the Indian Atomic Energy Commission; died in an air crash.

dark-skinned young man whom I took to be an Italian. After a while I pulled out a crime novel by Edgar Wallace in order to refresh my scanty knowledge of English, which I had learned after a fashion when I was twelve but never used since. The moment I did that the man opposite me said 'You are a physicist?' Surprised I asked 'Why should you think so?' He said 'You read Edgar Wallace.' Now surely it is quite wrong to assume that only physicists read Edgar Wallace; but he was right, I was a physicist. This confirmed my view that a really good scientist is one who knows how to draw correct conclusions from incorrect assumptions.

The man was Homi J. Bhabha, a handsome Indian from a wealthy Parsee family; he had studied in Cambridge and spoke impeccable English. When later I came to live in Copenhagen we became friends; he introduced me to Beethoven's late quartets, and I irritated him by complaining that his gramophone ran fast and played them in the wrong key. (Absolute pitch, valuable as it is

54

to a singer or string player, is a curse to a mere listener or a pianist faced with an out-of-tune piano.) Bhabha was also a very competent painter, and a first-rate theoretical physicist. But I was amused when one day he casually asked for instruction on how to use a Geiger counter; he was to travel to India by boat next week and wanted to measure the variation of cosmic rays with latitude. I told him the story of the boy who wanted to be a baker and was told he would have to serve a three years' apprenticeship: one year to run errands for the baker's wife; one to clean out the oven, and the last to learn how to bake bread. He smiled and got my point. Years later he became the head of the Atomic Energy Commission in India, only to die soon in an airliner that crashed into a mountainside.

The conference was a confusing affair, with so many people and so little time to sort them out. I was asked to give one of the talks, reporting on my work with Stern, and I remember how my chalk kept squeaking, whichever way I held it. Eventually Paul Ehrenfest, a famous Dutch physicist and one of the founders of the quantum theory, called out to me 'Frisch, if you don't learn to write without squeaking you will never be a professor.' I am ashamed to say that the simple trick for stopping chalk from squeaking didn't occur to me: all one has to do is to break it in half and thereby push the squeak into the ultrasonic region where it is no longer audible.

In Copenhagen I heard for the first time the suggestion that the fire that had devastated the German Parliament had not been started by the accused Communist, van der Lubbe, but had been deliberately laid by the Nazis in order to work up public opinion against the Communists; it was an idea that startled me at first but then seemed plausible. After my return to Hamburg my one remaining colleague in the department, Knauer, gave me dinner at his lodgings and wanted to hear what people abroad said about the fire. Although he had become a Nazi Knauer had never let the anti-semitic party line interfere with our friendship. He was slim and active, a member of a judo club, and had travelled in his canoe down the Elbe, all the way from Bohemia to Hamburg. He occasionally took me out in his canoe and on one occasion patiently paddled alongside while I swam across the Elbe, nearly a mile wide there. He still hadn't made up his mind what to call his canoe; perhaps 'atom' he said. When I suggested 'dipole' (something with opposite electric charges at the two ends) he blushed. But I found him married when I got back after the war.

Quizzed about the fire I tried to hold my peace and talk of other things. But when he insisted I did tell him that people were convinced that the fire had been laid by the Nazis for political reasons. He was horrified. 'But how can anybody think such a thing of people like Hitler or Goering; just look at their faces!' I personally didn't think that Hitler and Goering had faces to inspire such trust in their benevolence, but once again I held my peace and the matter was dropped. Knauer kept his friendly and helpful attitude to the last and made my departure easy by finding a small freighter that was going to London and had one cabin for a passenger.

On that cockleshell, on a windy day in October 1933, I left Germany with all my belongings in several trunks which kept sliding forth and back in my cabin as the ship rolled and pitched across the North Sea while I braced myself in my berth, unable to sleep. Once we had entered the Thames the ship quietened, and I could sit on deck and watch the flat landscape and then the dockland and the City of London until we finally tied up somewhere near Greenwich. I had to wait until the immigration officer came on board. When I showed him my passport he asked me 'Have you got a work permit? You must have one if you want to take a job in England.' 'I have no job' I replied 'I have a grant.' 'A grant is a high-class name for a job; you must have a work permit.' 'But how do I get one?' I asked him. 'You give the steward half-a-crown and send him ashore to phone your professor; see what he can do.' It worked like a charm; within two hours the immigration officer was back with his stamp, ready to let me in.

Nuclei

Atoms are not that small, about a thousand times smaller than microbes which you can see under an optical microscope. An 'ion microscope' shows quite clearly the beautiful regular pattern of atoms on the point of a sharp needle. But atomic nuclei are really small. Try to think of something a thousand times smaller than an atom and you are still not down to the size of atomic nuclei; you need another factor twenty or so. If an atom were enlarged to the size of a bus, the nucleus would be like the dot on this *i*.

So much depends on chance, even in science. Some scientist kept a box with photographic plates near an evacuated tube in which he produced electric discharges; when he found that the plates got fogged he kept them elsewhere and thus missed the discovery of X-rays. Konrad Wilhelm Röntgen in Würzburg was more observant; he realized that his electric discharges produced invisible rays that could penetrate opaque matter such as cardboard or human flesh, and after a feverish few weeks of investigation he published his startling results just before Christmas 1896.

It was the greatest sensation of the century; the newspapers seemed mainly concerned that the modesty of ladies might be violated by those indiscreet rays! Other physicists tried to find penetrating rays from sources other than electric discharges, and among many dubious claims that of Henri Becquerel in Paris stood firm: uranium and its compounds were 'radioactive', seemingly

15. Metal atoms forming part of a fine needle point, shown by the 'ion microscope'. The image is formed by gas atoms which, on striking the positively charged point, lose an electron and are then pushed away radially and spread over a photographic film of vastly greater area.

inexhaustible sources of rays which, like X-rays, could penetrate opaque matter, blacken photographic plates and let electricity pass through air.

Pierre and Marie Curie, also in Paris, confirmed all that but found that uranium ores were more radioactive than their content of uranium accounted for; so they searched for and found further radioactive substances in small traces (undetectable by ordinary chemistry) which they called polonium (in honour of Poland, the

16. Marie Curie (née Sklodowska) (1867–1934). Working with her French husband Pierre Curie, she discovered radium and polonium (named after her native country) in 1898. Her daughter and son-in-law, Irene and Frederick Joliot, discovered a world of 'artificially' radioactive substances in 1933.

home of Marie Curie, née Sklodowska) and radium. Others joined the search and within a few years found a good dozen further radioactive substances.

At the same time the radiations were seen to be of more than one kind. The most penetrating ones, resembling X-rays, went through thick metal plates and were called gamma rays; beta rays still passed through thin sheets of metal and moreover were deflected by magnetic fields, just like the electrons whose existence in electric discharges J. J. Thomson had established in 1897. The least penetrating ones, the alpha rays, were stopped by a mere couple of inches of air or by a sheet of paper; but Rutherford took a special interest in them because a magnetic field deflected them in the opposite direction. That proved that their electric charge was positive and that they were presumably flying atoms which had lost an electron or two. By 1908 Rutherford had definite evidence that alpha rays were helium atoms which had been flung out from a radioactive atom at several thousand miles a second and had lost two electrons each. But he didn't know yet that two electrons was all a helium atom had to lose, that alpha rays were just helium nuclei, travelling at several thousand miles a second.

All those rays made some chemicals glow in the dark; but alpha rays made zinc sulphide not just glow but sparkle, as was found in 1903 (in two places independently). It convinced the last doubters that atoms were real (just as spontaneous splashes on a calm lake may convince a sceptical fisherman). The thrill of being able to 'see atoms' led to the spinthariscope, a toy consisting of a strong magnifying lens, a layer of zinc sulphide and a trace of, say, polonium.

But to Rutherford it meant more: by counting the individual sparks ('scintillations') and thus the 'alpha particles' of which the alpha rays consisted he put the whole subject on a quantitative basis. He also studied the random deflections which the alpha particles suffered on going through thin metal foils; in that way he hoped to learn something about the structure of metal atoms. Most of the particles got deflected very little, and that agreed with current ideas about that structure. But a few were strongly deflected, some by more than ninety degrees. To Rutherford (so he said) it was as surprising as firing a canon ball at a piece of paper and seeing it bounce back.

For some days he kept muttering that there must be some incredibly strong forces somewhere in those atoms. They were thought to be diffuse lumps of positive electricity with electrons embedded like raisins in a pudding; not likely to let most alpha particles through with hardly any deflection but to fling a few of them back violently; electrons are very light and would simply be kicked out of the way.

With hindsight it seems simple. The atom is no pudding but mostly empty space with nearly all its mass in a very small volume at the centre. That 'nucleus' had to carry a positive electric charge to compensate for the negative charge of the electrons. Alpha particles – helium nuclei themselves – usually passed the nucleus of each atom at some distance and were only slightly pushed aside; that accounted for the small deflections. But occasionally an alpha particle would by sheer chance head almost straight for a nucleus and would come to feel the full electric repulsion which at close quarters could amount to a force of a pound or so, an incredibly large force for such a light particle, enough to stop it in its tracks and fling it back. Rutherford did a fairly simple calculation to work out how many alpha particles would be deflected by more than a given angle, and when the numbers were checked by counting the scintillations there was good agreement; the model fitted the facts.

I have already told how Niels Bohr applied Planck's quantum principle to the electrons surrounding the nucleus and how through the work of many people the structure of atoms was unravelled; that is, the way in which the electrons occupied that very large empty volume (compared to the nucleus) which we call 'atom'. As to the nucleus itself, Rutherford said at first that the exploration of its structure would be the job of the next generation; yet he was itching to have a go, and he had some clues.

An important clue came from the new radioactive substances. They were an embarrassment to the chemists: there were too many to fit into the well-established table of chemical elements; and there were some that appeared to have identical behaviour and could not be separated by chemical techniques. In 1913 a suspicion which had been growing was clearly spelled out by the English chemist Frederick Soddy, who created the new word 'isotopes' to indicate substances which were chemically identical but had different radioactive properties (or just different atomic weights). At the same time the physicists showed that some ordinary elements, for instance chlorine or copper, were mixtures of two or more species of atoms with different weights; that is, mixtures of two or more isotopes.

Those results gave startling support to a hundred-year-old idea that all the various atoms were tight clusters of hydrogen atoms, unbreakable by the tools of the chemist. That idea arose when atomic weights were first established and it was noticed that many of them were nearly whole numbers. But – regrettably – some were not, so the idea seemed wrong. Those exceptions were now seen to be mixtures of two or more isotopes; for instance chlorine turned out to contain two kinds of atom, with atomic weights of 35 and 37, mixed in the ratio of three to one; that explained the atomic weight of 35.5 the chemists had found. Such ideas had been aired quite early, but now there was evidence: streams of ions, made to fly at high speed through a vacuum, could be separated by magnetic fields into several streams according to their individual weights, which could in this way be measured. Those 'mass spectrographs' were soon improved to give very precise figures.

Of course it had been accepted since 1911 that practically the whole weight of an atom was contained in its nucleus; the lightest of them, the hydrogen nucleus, would then be the fundamental brick and was given a special name, the 'proton' (Greek for 'first'). So the atomic weight presumably was the number of protons in each nucleus of a given isotope. But that would give the nucleus a positive charge about twice as large as Rutherford's measurements had indicated; one had to assume that nuclei contained electrons as well, whose negative electric charge would partly counteract the positive charge of the protons. For instance of the two chlorine isotopes the light one would contain 35 protons and 18 electrons in each nucleus; the nuclei of the heavier one would consist of 37 protons and 20 electrons, resulting in the same charge of 17 units which gave both isotopes the chemical properties of chlorine, the seventeenth in the table of chemical elements.

Here was the beginning of a structure of nuclei; but so far their size was unknown, just as in the mid nineteenth century the chemists had a shrewd idea of the structure of some molecules but none of their size. Rutherford had good evidence that his alpha particles had been turned back purely by the electric repulsion of the nuclei in his gold and silver foils without ever touching the nuclei themselves, which therefore must be below a certain size. But how much below he couldn't tell.

Then the 1914–18 War intervened. German submarines tried to starve England into surrender, and Rutherford, together with others, spent his time developing underwater listening devices. Only towards the end he managed to get back to his beloved alpha particles and to show that when they passed through nitrogen gas he occasionally observed a fast proton. Apparently the much weaker repulsion of a nitrogen nucleus (compared with one of gold or silver) could not prevent an occasional collision. Here was evidence that nitrogen nuclei, at least, contained protons; and the frequency of collisions gave some idea of the size of the nuclei. Moreover the protons had much more penetrating power than the alpha particles that released them. It was as if a slight tap had released a coiled spring. 'The atom has been split' said the newspapers; energy had been released from atomic nuclei. It was calculated that the Queen Mary could sail several times across the Atlantic on one ounce of nitrogen if the energy of its nuclei could be utilized.

There was only one snag: one can't aim alpha particles at nitrogen nuclei, and most of them missed, wasting their energy by having to push electrons out of the way, and turning into useless helium atoms. It was, as Einstein characteristically put it, 'like shooting sparrows in the dark'. Each hit liberated a few times the energy of the bullet, but what use was that when millions of bullets had to be wasted to score one hit?

All the same, it was an important new clue, and physicists in other places started to repeat and extend Rutherford's experiment. As expected, only light nuclei produced protons under alpha-ray bombardment; with heavy ones the electric repulsion was so strong that it prevented collisions. The technique was simple: one needed a source of alpha particles, for instance a thin layer of polonium (ideal because it emits hardly any other rays), a layer of the element one wished to bombard (or of a suitable compound), a layer of zinc sulphide with a strong magnifying lens to observe the scintillations, and some mica or metal foils to insert in the path of the particles that caused them, for measuring their penetrating power.

But scintillations are hard to see. First one must sit in total

17. John D. Cockcroft (1897–1967) and George Gamov (1904–1968) about 1930. The Englishman who first (in 1932, with Walton, joint Nobel Prize in 1951) split nuclei with artificially accelerated protons, and the Russian theoretician who persuaded him that it was worth trying. Gamov soon settled in the U.S.A.; Cockcroft was knighted in 1948.

darkness for half an hour so that one's eyes become fully sensitive; even then the eye will miss some scintillations if they come too frequently – say, more than fifty a minute – and 'invent' some if there are fewer than one or two a minute. At that rate it takes a long time to get good statistics, and subjective errors can be serious. I don't know how I escaped the fate of many physics students in Vienna who had to serve shifts, counting scintillations, in darkness mental as well as physical: to secure unbiassed results they were not told what they were counting! It didn't work; they sensed the satisfaction of their supervisors on having large numbers of scintillations reported, and in the hectic atmosphere of 'we can do better than the English', papers were published which for a while made Vienna the *enfant terrible* of nuclear physics. An objective, fast and reliable way of counting particles was clearly needed.

The means were at hand: radio valves had been developed (partly because of military interest during the 1914–18 War) and were being gradually improved. In the early twenties they still produced a lot of 'noise' (irregular output) and didn't insulate well enough, but by 1930 selected valves were good enough to amplify the tiny pulses of electric current which occurred whenever an alpha particle or a proton passed through the air between two electrodes (metal pieces with opposite electric charges) in an 'ion chamber'. Air (as all gases) is a very good insulator; but an alpha particle or a fast proton, kicking electrons out of its way, leaves a trail of ions – electrically charged atoms and molecules – which are attracted to the electrodes, producing a brief pulse of electric current. Amplified by valves, they could be counted on a meter (similar to the one in a taxi).

Those improvements in technique were decisive. By the mid-thirties, counting rates of thousands a minute were commonplace, and what would have taken years could now be done in weeks. Moreover from the size of the pulses in such an ion chamber the number of ions formed by each particle could be estimated, and hence its energy.

At first it was not possible to count beta rays (fast electrons). They also leave a trail of ions, but it is less dense and much longer, and the scintillations they cause are too diffuse to be seen. Hans Geiger had worked on electric counters as early as 1909 in Rutherford's laboratory. Having returned to Germany, he developed the 'Geiger counter' which produces and records a small electric discharge even if a particle in passing through it creates only a few ions. It was a simple device, a pumped-out tube with a positively charged wire along its axis, soon to be found in many laboratories. To the public it became the symbol of the 'atomic' (that is, nuclear) physicist, just as the quill is the symbol of the writer though most writers type or use a tape recorder. Nowadays the Geiger counter, though a few may survive in school laboratories, is as dead as the quill. Today we count the scintillations produced by beta rays with sensitive photocells, even more rapidly and reliably. But I am getting ahead of my story.

In the meantime it was noted with surprise that the nuclei of some light elements, in particular beryllium – a light, hard, silvery metal – could apparently not be split; at least they produced no protons when bombarded with alpha particles. Instead they produced, as Walther Bothe in Heidelberg found in 1930, a weak but very penetrating radiation, presumably 'hard' gamma rays, that is, X-rays of very short wavelength and hence very high quantum energy. But when in Paris, Irene (Marie Curie's daughter) and her husband Frederic Joliot measured the power of various substances to stop those mysterious beryllium rays they were surprised to find that wax had the opposite power: on passing through wax the rays appeared to get stronger!

The Curie–Joliot couple guessed the reason. Wax contains a lot of hydrogen, and apparently the rays were able to kick hydrogen nuclei (protons) into rapid motion; those fast protons, emerging from the wax, made more ions in the air so that the radiation seemed stronger. With a Wilson chamber the Paris team showed that fast protons were indeed coming out of the wax; but they still thought that the 'beryllium rays' were gamma rays. The theoreticians who said the explanation must be wrong got little attention in the Curie Laboratory.

But in the Cavendish Laboratory in Cambridge where Rutherford had been since 1919 the Paris results caused great excitement. Could this be the 'neutron' – the proton without electric charge – about which Rutherford had speculated as early as 1920? Some unsuccessful attempts had been made to find such a particle. It was expected to go through inches of solid matter because, having no electric charge, it would disregard the electrons in its way and be stopped only by eventually running slap into a nucleus; just what the beryllium 'rays' appeared to do!

Thus in Cambridge the physicists were on the lookout for a neutron, and the right tools were available: ion chambers with which James Chadwick could measure how many ions various nuclei (such as those of hydrogen and oxygen) created after being kicked into motion by the beryllium rays. From that he could prove that those 'rays' were really fast particles that weighed about the same as a proton but had no electric charge.

I was told that Rutherford later met Joliot and asked him 'Did you not realize that you had those neutrons which I discussed in my Bakerian lecture in 1920?' and that Joliot replied 'I never read your lecture; I thought it would be the usual display of oratory, not of new ideas.' How wrong he was! But he got his own back when, just two years later, he and his wife discovered artificial radioactivity, which Rutherford had missed (as he admitted) by looking in the wrong direction.

The discovery that neutrons could be knocked out of nuclei made it seem likely that they were present inside them. Indeed there was no longer any need for the assumption (which had been difficult to reconcile with the quantum theory) that nuclei contained electrons. A nucleus, say cf the isotope chlorine-35 could consist of 17 protons and 18 neutrons; a chlorine-37 nucleus would just contain two more neutrons. That simple model is still universally accepted.

But what about the beta rays, the fast electrons which were sent out by some radioactive substances? Did they not prove that electrons were present in some atomic nuclei? Perhaps a neutron was – as Rutherford had surmised – a close union of a proton and an electron? That idea seemed plausible until, later in 1932, it was found that positively charged electrons – soon to be called positrons – existed as well. So perhaps the proton was a close union of a neutron and a positron? We can't have it both ways!

Heisenberg suggested that we should regard the proton and the neutron as two different states of the same entity, which came to be called a *nucleon*. The beta-emitting nuclei found in nature contain too many neutrons – or, if you like, too many nucleons in the neutral state – so they tend to turn one of them into a proton to improve the balance. Of course the extra unit of positive electric charge cannot just come out of nothing; it must be balanced by an equal negative charge which is flung out from the nucleus as a beta electron. The 'artificially' radioactive nuclei discovered in 1934 had too few rather than too many neutrons; they restore the balance by turning one of their protons into a neutron and creating a positron to compensate for the loss of electric charge.

The discovery of the positron is an intriguing story. In 1911 the Scotsman Charles T. R. Wilson, interested in the minute droplets that form his native fogs and clouds, found that ions (electrically charged atoms and molecules) tended to attract water and thus form droplets. By suddenly expanding a container with moist air in it he could, for a fraction of a second, create air super-saturated with moisture which would condense as a water droplet, wherever an ion was present. When during that split second a fast alpha or beta particle went through such a 'cloud chamber' (or Wilson chamber) the trail of ions it formed became visible as a trail of droplets.

At first this looked again like becoming just a nine days' wonder, a way of making those fast submicroscopic particles visible, rather as a high-flying aeroplane shows up by its vapour trail. The technique was tricky and after the first excitement was neglected for a decade. But gradually, as more physicists mastered the precautions needed to make it work reliably, the cloud chamber found its place. While large numbers of nuclear particles could be counted with scintillation screens, ion chambers and Geiger counters, it was a bit like a game of Twenty Questions; you keep getting answers of yes or no, and to piece together meaningful information you have to ask a great many questions. But with a cloud chamber a single successful expansion gives you a snapshot of what had happened in the chamber at that instant, with revealing and often surprising detail, well worth the trouble of building a fairly complex and delicate apparatus and of having to wait a minute or more after each expansion to let the disturbance settle down. More and more, it was Wilson's cloud chamber that could tell you what was going on, whereas counting devices were used, once the events were understood, to accumulate the large numbers required for accurate statistics.

Early this century it was realized that some ions were formed in air even in the absence of any known radiations, and the Austrian Victor Hess, taking his equipment up in a balloon, found that more ions were formed at higher altitudes. Evidence gradually grew that there was a 'cosmic radiation' coming from outer space, day and night. In the late twenties, cloud chambers showed an occasional track, apparently caused by a very fast electron. By placing the chamber in a magnetic field one got bent tracks because the particles that made them were deflected by the field. The direction of bending indicated whether the particle was positively or negatively charged, and with some the charge seemed to be positive; but they might have been ordinary electrons travelling in the opposite direction, which the track does not tell us.

In 1933 Carl D. Anderson in California photographed the first track of an electron whose charge was undoubtedly positive; he could tell because a lead plate he had put in his cloud chamber had diminished the energy of the electron that had gone through it so that the track afterwards was more curved; that left no doubt which way the electron had travelled. Soon afterwards Patrick M. S. Blackett in London got photographs which showed a bundle of tracks, some bent to the right, some to the left, clear evidence of the simultaneous emergence of positive and negative electrons from the chamber wall.

The surprising thing is not that positrons were found but that they hadn't been identified earlier: they had actually been photographed before, but in each separate instance the evidence had been regarded as unconvincing. And yet the existence of positrons had actually been predicted from Dirac's theory as early as 1928; but most experimenters paid little attention to such 'mathematical speculations'. We experimental physicists really are a cautious lot, hard to convince of what is under our noses!

London 1933–1934

So here I was in England, the land which Goethe had so much admired that I expected it to be inhabited almost entirely by supermen. Some features appeared to confirm that expectation. I remember greatly admiring the roadworkers who invariably put their jackets on before biting into their sandwiches; it agreed with what I had heard about the habit of 'dressing for dinner'. (The true reason was probably that it was October, and they were cold.) My boss was both impressive and charming: a tall man with vigorous features who greeted me with a handshake every morning I came to the laboratory. It took me weeks to discover that as a rule the English didn't do that; Blackett wanted to make me feel at home. He later became President of the Royal Society and died as Lord Blackett; in 1933 he was head of the physics department in Birkbeck College, housed in a tall old building near Fetter Lane in the legal quarter of London. It was a college for working men, attending evening classes, and during the day the physics department, with laboratories both on the top floor and in the basement, was given over entirely to research. I learned to run up those four floors in thirty-five seconds, and down in twenty-five.

There were hardly any English people in Blackett's crew, indeed hardly any two people of the same nationality; we jokingly called ourselves the League of Nations. We were not too impressed with English cookery, in particular with the cheap restaurants from

18. Patrick Maynard Stuart Blackett (1897–1974). British naval officer who turned to physics (Nobel Prize 1948), pupil of Rutherford, later initiated the study of rock magnetism. President of the Royal Society in 1965, raised to the peerage in 1969. A born leader and strategist.

which a strong smell of boiled mutton used to emanate: though I wouldn't go as far as my friend Fritz Houtermans who stated in public that the British were a poor nation who lived on the residues of wool manufacture! But we preferred French and Italian restaurants, and we used to go there in a body, a dozen or so of us; when we parted afterwards in the street I sometimes felt that probably the natives rented balconies in order to watch the extraordinary spectacle of a dozen people exchanging sixty-six handshakes before departing on their several ways. We all were busy learning English, and I remember that on one occasion about half a year after we had arrived, we tested our vocabulary. We got hold of the *Shorter Oxford Dictionary* and picked pages at random, read out all the words on a page and checked how many of them we could define. Much to my surprise we all of us found that we could define about half the words in the dictionary, which contained about 50 000 words. We were gratified to deduce that we each of us had mastered about 25 000 words of the English language; not a bad record after six months, we thought.

Fritz Houtermans I had met in Berlin, but in London I saw a lot more of that impressive eagle of a man, half Jewish as well as a Communist who had narrowly escaped the Gestapo. His father

71

19. Fritz Houtermans (1903–1965). The maverick physicist of Dutch–Austrian parentage who never achieved the success which his originality deserved. (Caricature by the author.)

had been a Dutchman, but he was very proud of his mother's Jewish origin and liable to counter anti-semitic remarks by retorting 'When your ancestors were still living in the trees mine were already forging cheques!' He was full of brilliant ideas, with a profound understanding of quantum theory, one of the first to apply it to atomic nuclei.

In England he worked at the research laboratory of His Master's Voice and tried to verify a prediction made by Einstein in 1909 that a light beam could become stronger rather than weaker on passing through a gas containing the right kind of excited atoms. But he probably forgot to bring his apparatus roses to make it work smoothly (he said he did that with important experiments); an expensive transformer burnt out, and his boss would not replace it. If his experiment had succeeded the laser might have been invented twenty years earlier.

After a year in England he achieved his goal of going to Russia, but fell victim to one of Stalin's purges and spent a couple of years in prison; his wife with two small children managed to escape and get to the U.S.A. When Hitler made his temporary pact with Stalin in 1939 it included an exchange of prisoners, and Houtermans was handed back to the Gestapo, but freed through the intervention of the Nobel Prize winner Max von Laue, one of the few German scientists with the prestige and courage to stand up against the Nazis. After working for a time with the successful inventor Manfred von Ardenne he even got back into academic life. Later he got a professorship at the University of Berne; how the stolid Swiss could get on with such an eccentric character I'll never understand.

He had been a chain smoker, and lung cancer was spotted when a fairly trivial accident put him into hospital. He was only sixty-two when he died.

The biggest piece of equipment on the top floor of Blackett's laboratory, and in some ways the centre of his work, was the triggered cloud chamber which he had invented and built. There was much interest in the 'cosmic radiation', known to contain very fast particles; but it was only very rarely that an expansion showed any of them. Now Blackett had invented a way whereby a particle would take its own picture: on passing through the chamber it would also go through one or more Geiger counters which would cause its photograph to be taken, by triggering the expansion of the chamber. That trick enormously increased the number of photographs that showed interesting events and led Blackett and his collaborators to many important discoveries.

The chamber stood in the middle of the laboratory, an object a good yard or so in size, and the alarming thing was that you never knew when a cosmic particle would trigger it and make it go off with a bang like a cannon. So we all did our work as best we could, nervously expecting to be startled out of our wits at any time during the next hour or so. When that happened the Swiss member of our 'League of Nations', Gerhard Herzog, rushed up, removed the photographic plate which had taken the picture of whatever had triggered the chamber and handed the plate to his young wife, who had volunteered to help and disappeared with it into the darkroom.

She appeared again about ten minutes later, when we all crowded round to see what fish we had caught. On one occasion she didn't reappear for quite a while, and nobody noticed until suddenly her husband said 'What's happened to her? I gave her that plate twenty minutes ago!' and rushed into the darkroom expecting that she had fainted. They both came back a moment later, his wife wearing a huge grin and teasing him for the rest of the day with the repeated remark 'What a husband! Twenty minutes he has let me die in there!'

Once Blackett gave two tickets to Herzog and myself for the Friday evening lecture by Lord Rutherford in the Royal Institution. We worked until it was time to go and then went there in the shabby brown lounge suits which we usually wore in the laboratory. On arrival we saw gentlemen in tail coats and ladies in long dresses stepping out of limousines and walking into the brightly lit foyer. We watched that spectacle for a while, then Herzog said 'You can do what you like, I won't go in.' I was a little more courageous; I went up to an usher and asked 'Can I go in like this?' The usher looked me up and down and said 'If *you* don't mind!' I have hardly ever heard such a brief and yet so illuminating remark; for the first time I realized that wearing evening dress was not a duty but a privilege! We did go in and enormously enjoyed the lecture in which Rutherford with great gusto demonstrated atom splitting with the first high-voltage set, built in Cambridge and specially taken to London; there was another like us, trying to hide behind a pillar.

The particular task that Blackett had set me – to find the gamma rays which were expected to result from the mutual annihilation of positrons and electrons – never came off. We were beaten not by one but by two Frenchmen, who immediately started a quarrel about various details. I was glad to be out of it, and started on a project of my own: to construct a cloud chamber which would be sensitive not merely for about one-tenth of a second but a much longer time – a second or more – so that one could see the tracks appearing one by one and perhaps make useful observations without the complex system of Blackett's triggered chamber. By expanding the air in the chamber rather slowly I achieved a sensitive time of well over a second, and the chamber was very pretty to watch and easy to make; but it was no use for research and never caught on as a demonstration instrument. Then I might have been at a loose end but for the great windfall: the French discovered 'artificial radioactivity', and I managed to be one of the first to climb on the band-waggon.

In 1933 it was generally believed that only a dozen or so radioactive elements (that is, elements with unstable atomic nuclei) existed, and that they were all known. But in January of 1934 Frederic Joliot and his wife Irene, a daughter of Marie Curie, were studying the positrons which they had found to be emitted from aluminium when it was bombarded with alpha rays. They used a Geiger counter to observe those positrons and kept complaining that the counter misbehaved: it still went on counting for a few minutes, at a decreasing rate, when you removed the source of alpha rays. In vain they replaced the counter; but when the aluminium was removed the counting stopped at once. It wasn't the counter's fault at all: it was that the aluminium emitted positrons not immediately on being struck by an alpha particle but with a random delay of several minutes. In other words, the impact of an alpha particle turned an aluminium nucleus into a radioactive nucleus which had a mean life of a couple of minutes.

Joliot and Curie published their observation in the middle of January 1934, and the excitement was immense. Clearly there must be many other cases where the impact of an alpha particle might turn an ordinary stable nucleus into an unstable one, and a number of people instantly began bombarding various elements with alpha particles. I felt I had to go one better; in order to spot very short-lived products I built an apparatus (from pieces bought at Woolworth's) which could whip a sample in a fraction of a second from the alpha source to a nearby counter, well protected from its gamma rays.

Within days I had results with phosphorus and sodium (neither of them so short-lived that my device was needed!) and Blackett made sure that the work was completed and published quickly. As soon as I had written a brief note about my results he telephoned the editor of *Nature* and at the same time sent me along with my manuscript. As a result the paper got published within six days, which I think may well be a world record.

It must have been about that time that Niels Bohr came to visit Blackett. I had of course seen him in Copenhagen, but there were so many people that we had hardly exchanged more than a few words. This time Blackett probably pointed out to Bohr that my grant would run out in October, and persuaded him that I would be useful in Copenhagen. That is sheer guesswork; all I know is that Bohr came to talk to me and took me by one of my waistcoat buttons and said 'You must come to Copenhagen to work with us. We like people who can actually perform thought experiments!' He was alluding to my last experiment done in Hamburg when I was working with Stern; it had previously been discussed as if it was possible only in thought. Anyhow I felt very bucked up by Bohr's visit, his kind remark and his immensely impressive and yet benevolent face, so that I wrote to my mother 'You need no longer worry about me; God Almighty himself has taken me by my waistcoat button and spoken kindly to me.'

As I continued to live in London my initial exaggerated respect for the English began to turn into the very opposite. For instance, the London thoroughfares, like Oxford Street, were so narrow that overtaking was impossible; long rows of motor-cars proceeded at a funereal pace, controlled by the speed of the occasional horse-drawn van. And how could the inhabitants of a metropolis put up with a bus system of about a dozen bus companies; with buses of all shapes, sizes and colours, many with their top deck open to the elements; and with stop signs that didn't indicate which of the bus lines would stop and which wouldn't. I was saved a lot of trouble by going about with a colleague, Werner Ehrenberg (the German member of our 'League of Nations') who limped quite badly; he only had to lift his cane, and every bus would stop to let him get on, and me with him.

Having lived in Germany for six years I had accepted as obvious that one had to be consistent and thorough; it took me a while to comprehend that muddle had its virtues, in particular when it was softened by the humanity of which the kindness of bus drivers to my crippled friend was a good example. I gradually came to see that every human quality, like a medal, has two sides; that German consistency and thoroughness had 'rigidity and ruthlessness' stamped on its obverse, and that the muddle and lack of organization in England meant flexibility and humanity as well.

Another thing that puzzled me at first was the fact that, as far as I could see, my English colleagues knew very little physics outside their narrow speciality. On the other hand they knew a lot more about politics, literature and sports than their German counterparts. Once again I gradually learned something: I came to realize that the thorough knowledge of a speciality like physics (or any other) without some understanding of politics and human behaviour did little to protect the average German from a vicious ideology like National Socialism. Birkbeck College once announced an evening lecture by John Buchan, with the title 'Margins of Life', and I expected the speaker (whose name was unknown to me then) to talk about some scientific field, perhaps about viruses or very large molecules on the borderline between inorganic matter and living organisms. Not a bit: what he spoke about, to my growing astonishment, was the importance for a student not to work too hard! A student should not devote his entire time to the study of his subject; he should leave a margin on which he could scribble notes on what went on around him. I was quite amazed that such advice should be regarded as necessary; I felt that students were generally a scatter-brained lot and in my view ought to be encouraged to stick to their books. But the lecturer obviously thought that the opposite advice was necessary to prevent them from becoming narrow-minded.

I didn't see much of England during this one year's stay. I

remember taking a train to Southampton once and going for a two days' walk, spending the night at an inn. It was very pleasant but I didn't really know what to look for, and all I saw was a flattish landscape with some agreeable cliffs here and there. Later in the summer I decided to go further afield and booked a room with a landlady in Windermere to make the acquaintance of the famous Lake District. It turned out to be very much like the Austrian Lake District, the Salzkammergut, famous for its persistent rainfall. After four days I went back to London because the rain never stopped. But my landlady insisted that I had booked for a week, and a week's bill for bed and board I had to pay.

The working conditions at Birkbeck College also gave me a shock. Of course I had been spoiled at Stern's institute where two first-rate mechanics and a good glass-blower were at our disposal, as well as a supply of up-to-date instruments and materials. I remember writing home from England that to build equipment was so much patchwork and make-do that after a few weeks of it my imagination boggled at the thought of asking for a piece of rubber tube 18 inches long! What saved me was the existence of Woolworth's. In those days no item cost more than sixpence. Admittedly a pair of socks cost a shilling; but Houtermans once insisted on buying one sock, explaining that it was a present for somebody who had only one leg. One could buy almost anything there. Once I bought a piece of ladies' black underwear; it was the easiest way of getting hold of some smooth black fabric for lining my cloud chamber. I didn't have the courage to charge the laboratory for that particular purchase.

The workshop consisted of two young mechanics who largely cancelled each other out by chatting instead of working; and the head was an elderly mechanic whose main job in life was preventing the scientists from using up too many screws. Whenever we wanted any he would rummage in a drawer containing hundreds of screws, mostly bent and rusty. Altogether the Whitworth system of classifying screws defeated me; it could only be due to the impulsive action of a nineteenth-century manufacturer who got so fed up with the muddle that one day he ordered all the screws in his workshop to be collected and numbered according to size. Another thing that bothered the chief mechanic – probably with reason – was the borrowing of tools. He had hung up a large sheet of paper with the words 'PLEASE RETURN TOOLS AFTER USE' written on it, and our 'League of Nations' wrote translations underneath in whichever language the writer was familiar with; in a short time it was written there in a dozen different languages. When I came back, some five years later, I found that the list had swelled to several dozen languages, many in incomprehensible characters like Urdu or Arabic. But after the war, when I asked about that sheet I was told 'Oh, it got so dirty, we threw it away.' I still think that was a shame.

I also noticed that most of the people I talked to – other than foreigners – did not speak English. They spoke Cockney. When I asked a lady at one of the occasional parties to which I got invited where I might take lessons in Cockney she was horrified; I couldn't see why. I had the feeling that I got invited largely because one had to be kind to foreigners; after all it wasn't their fault, poor dears, that they had the misfortune of being born outside the British Isles. But I was quite surprised at the amount of spontaneous kindness that I found, in particular about the habit of complete strangers to talk to me in a railway carriage. The conversation always started with some remark about the weather, and Houtermans insisted that until quite recently Britain apparently didn't have such a thing as weather, otherwise it wouldn't be such a popular topic of conversation. It took me a while to realize that starting a conversation with some remark about the weather was a very good social invention; the person so addressed could either break off the conversation by some equally trivial reply, or lead it to anything from cricket to politics or theatre.

During that year I laid no more than a tenuous basis for an understanding of England. Perhaps the only thing I learned was that the German way of life wasn't the only one. I knew I would be going to Denmark and was excited though I didn't know what to expect. Anyhow I felt that my stay in England had been just an interlude, and I concentrated on getting my work wound up and my belongings packed, and then I was off.

Once again I had to cross the North Sea, but this time I took the ferry-boat that travelled regularly between Harwich and Esbjerg. Instead of a gale I encountered fog, and we had to wait several hours on a dead calm sea before the captain dared to nose his way into Esbjerg harbour (it was in the days before radar!). And from there by railway and two ferries to Copenhagen.

Denmark 1934–1939: 1

Niels Bohr's institute was international in quite a different way. It was the Mecca of the world's theoretical physicists. There were many foreigners about, but they kept changing; most of them were transient visitors who would come and give a seminar, a talk or two, and disappear again. One of the first talks I heard was given by George Gamov. I cautiously enquired what language the famous Russian physicist was going to speak and was told 'Danish; but don't worry, you'll understand him.' How could I, having been in Denmark only a few days? I hadn't even started taking Danish lessons. But all the same I understood Gamov; he peppered his Danish with English and German words, gesticulated and made funny drawings. He really knew how to communicate and was very entertaining, as anybody will understand who knows his 'Mr Tompkins' books in which the mysteries of physics are expounded to laymen in a very amusing if not always quite accurate way.

Another person with whom I became closely acquainted was George Placzek, a Bohemian in every sense of the word. He came from what later became known as Czechoslovakia, had studied in Vienna and been all over Europe. I reckon that by the time I left Copenhagen he spoke ten languages more or less fluently and with a fine range of naughty verse in most of them. As I met him he had just accepted an appointment as Professor of Theoretical Physics at the new University of Jerusalem and was due to leave shortly.

20. George Placzek (1905–1955) the versatile Bohemian who contributed to the theory both of molecules and of atomic nuclei. (Sketch by the author.)

To pack his belongings was a tiresome chore, and he asked me to keep him company to prevent him from falling asleep. He kept up a barrage of talk and comment while I sat there and idly thumbed through the books he hadn't packed yet. At one point he gave a sudden shout and held up a piece of paper: the receipt for a sum of 100 German marks which he had left as a deposit with the Berlin University Library, a deposit which he had quite forgotten but would now be able to recover on passing through Berlin. That happy windfall had to be celebrated; tooth glasses were found, and brandy was drunk from them. After the ceremony, the piece of paper had of course disappeared and Placzek started to unpack again with many curses and imprecations. Finally he found it again and held it under my nose so that if it disappeared once more at least there should be a witness that the whole thing hadn't been a dream.

The packing went on for a couple of days, with slowly increasing frenzy as his departure drew near. On the last evening we were both trying to stuff the rest of his belongings into a trunk, and I vividly remember a large eiderdown which kept extruding one corner whenever the other three had been squeezed into the trunk. In the middle of this he suddenly left me to it and started dictating a letter in Danish to one of his friends who had turned up to help. That letter contained detailed instructions to the carriers, what to do with his various belongings, which things to ship to Israel and which to put in store, either to be forwarded later or to be kept in case he returned to Denmark.

Ten minutes before the train was due to leave we all raced down the stairs with trunk and suitcases and crammed into a taxi, telling the driver to go hell for leather to the station. During that drive Placzek implored Divinity to let him catch the train, with promises that he would be good in future and never arrive at the railway station less than ten minutes before the train was due. We arrived one minute *after* the scheduled departure; Placzek ran behind a panting porter who had grabbed his cases. Hopeless it seemed; but the train was still there! The sleeping-car attendant had noticed that one of his flock was missing and had held the train for two minutes; so Placzek caught it.

We later had numerous communications from Israel; how at first he had to fend off the persistent requests from the univerity authorities that he should give his lectures in Hebrew, which he had not yet learned. They gave him a year to learn Hebrew. (He learned Arabic as well.) When at the end of that year he still wouldn't give lectures in Hebrew – he felt the language couldn't cope with modern physics – and they insisted on it, there was a telegram 'THROUGH WITH JEWS FOR EVER', and he came back to Denmark.

The scientific language universally used in the laboratory was English; the Danes, by and large, understood it, and certainly the scientists did. All the same it was clear that I would have to learn Danish, and I found a teacher, a charming old lady of eighty who taught me three times a week, with great patience and clarity. Within a few weeks I had acquired a spurious fluency which caused one of my friends to say 'Frisch really knows only twenty words of Danish, but he uses them as if he knew a great many more'.

I gave my first Danish seminar talk four months after my arrival, and it went reasonably well until the questions came; although I could *speak* Danish after a fashion I found it almost impossible to *understand*. Even after five years, when I could converse with ease, I often couldn't guess what two students were chatting about.

I believe that the Danish language developed from shouts exchanged between fishermen over the breakers along the coast, where consonants are inaudible. Danish, casually spoken, appeared to contain no consonants at all but just a sort of thick soup of vowels interrupted by an occasional glottal stop. But do not think that I despise the Danish language. It is flexible and simple and has many telling phrases I have not encountered in other languages. If it is clearly pronounced it is one of the best and most expressive languages that I know. Though it is of course closely related to Norwegian and Swedish, the other two main Scandinavian languages, it is much softer and sloppier, and I always felt that the Danes were the Austrians of Scandinavia.

The Institute of Theoretical Physics in those days had a flat roof on which we often walked about and argued; a sunny place with a lovely view over the park behind the institute. (It was later roofed over and turned into a canteen.) One day, to our amazement, we saw Placzek climbing out of a small window and dropping onto the roof with a fiendish grin; he had bolted the toilet behind him. A proposal to push him back through the window was abandoned as impractical, and for several days we had to go down one floor to use the toilet there, which was a nuisance. In the end Felix Bloch, a Swiss mountaineer as well as a brilliant theoretical physicist, squeezed his hefty frame with many curses through the window and unbolted the convenience.

On another occasion Placzek suffered an unexpected public defeat. We were all walking along the water which surrounds the centre of Copenhagen, a semicircle of artificial lakes constructed a couple of centuries ago when the swamps around Copenhagen were drained, and we grumbled about the nuisance of having to

21. Hendrik Casimir. Dutch student of Niels Bohr, who later became director of the Philips (Eindhoven) research laboratory and president of the European Physical Society.

walk for perhaps a quarter of a mile to the next causeway where we could cross over. When Placzek jokingly said 'One really ought to swim across' the challenge was taken up by Hendrik Casimir, a young Dutchman who later became Director of the Research Laboratory of Philips, Eindhoven. Casimir offered to swim across for 20 kroner, a sum approximately equivalent to £1 in those days, and Placzek bet he wouldn't; the presence of Mrs Casimir gave him confidence. Casimir accepted the bet, solemnly took off his coat and handed it to his wife, who took it without a word of protest, waded into the water and swam across. We all went to the nearest causeway and met him on the other side. There he accepted his coat back, put it on, held out his hand for the 20 kroner and spent a good part of it on a taxi to get to his hotel. The next day he went back to Holland after his brief visit; when we saw him off at the railway station we noticed with amusement that he was wearing his dinner suit, obviously his only dry outfit.

Why is it that scientists are liable to waste their time with such childish pranks? These were all grown-up men, men in their late twenties, with a considerable reputation for scientific achievement. Then why this schoolboy behaviour? Well, I think scientists have one thing in common with children: curiosity. To be a good scientist you must have kept this trait of childhood, and perhaps it is not easy to retain just one trait. A scientist *has* to be curious like a child; perhaps one can understand that there are other childish features he hasn't grown out of.

At first my work was not very exciting; I continued the kind of experiments which I had been doing with Blackett in London, looking for new radioactive elements produced by alpha-ray bombardment, and I was lucky in finding two more to study. Otherwise I tried to make myself useful; I improved amplifiers and Geiger counters, which were then used by other scientists in their research on radioactive materials. Some of those counters were deliberately made from thin-walled metal tubes, turned even thinner on the lathe and barely capable of resisting the pressure of the air when they were evacuated to be used as counters. This led to one incident which was amusing, at least to the onlookers. Niels Bohr had done experimental work in his young days and often seemed to regret that he no longer found time for that, being wholly engaged in theoretical physics and administration. He loved to come into the laboratory and watch us. One day he said 'Can't I be of some use? I'm not as clumsy as I look', and before we could warn him he picked up one of the thin-walled counters, which immediately crumpled up with a nasty crackling noise. Niels Bohr dropped it as if it had burned his fingers, and tiptoed out of the room very embarrassed. But how could he have known that such a counter would collapse under the gentlest touch?

The year 1934 had been a memorable one for physics. It had opened with a fanfare, the discovery of artificial radioactivity, and many of us (including myself) had jumped onto that band-waggon. Within a couple of weeks someone who had been to Italy told me in London that Enrico Fermi was preparing to bombard elements,

22. Enrico Fermi (1901–1954). The Italian pioneer in both quantum theory and neutron physics; after his Nobel Prize in 1938 he settled in the U.S.A. and directed the construction of the first nuclear reactor. His school revitalized Italian physics.

not with alpha particles (as we all did) but with neutrons. That puzzled me, since neutrons were very rare: you had to bombard beryllium and waste a hundred thousand alpha particles to produce one neutron. What was the sense of using such expensive bullets? I had not seen at once (as Fermi had) that those expensive bullets were also extremely effective: a neutron was sure to hit a nucleus eventually, being neither repelled by its positive electric charge nor bothered by the electrons which, in solid matter, caused an alpha particle to stop within a fraction of a millimetre, with very little chance of hitting a nucleus on that short journey.

As soon as the neutron was discovered in 1932, Fermi with his strategic sense had realized that experiments with that new brick of matter must have first priority, and had started to assemble a team of young physicists, some of whom he sent abroad to become familiar with various techniques. (One of them, Emilio Segrè, had come to Hamburg, and we had worked together for a while.) When artificial radioactivity was discovered Fermi was ready, and his first paper appeared within a month. It was sensational: practically all the elements he had bombarded were rendered radioactive; some yielded two or more different active products, each with its characteristic rate of decay.

A light nucleus, struck by a neutron, would often eject an alpha particle (helium nucleus) or a proton (hydrogen nucleus), thus becoming transformed into a nucleus with a different (lower) electric charge and hence different chemical nature; those – usually radioactive – nuclei could easily be separated from the material bombarded, by the chemist in Fermi's team. But heavier nuclei invariably just swallowed the neutron, thus turning into a heavier variety (a heavier isotope) of the original nucleus, chemically indistinguishable from the original but often betrayed by its radioactivity.

At first Fermi used whatever table was handy in placing his neutron source (an inch-size capsule containing a mixture of a radium compound and beryllium powder) close to whatever he wanted to bombard; but the results were erratic, and a wooden table produced stronger radioactivity than one of metal or stone. He guessed that this was due to the hydrogen contained in wood; a neutron bouncing back from a hydrogen nucleus – like a billiard ball recoiling from another – would have lost much of its speed, and perhaps slow neutrons were more effective. So he bombarded various elements surrounded with paraffin (rich in hydrogen) and was amazed to find that the radioactivity of some of his samples was increased a hundredfold!

Fermi had made an arrangement with *La Ricerca Scientifica* to send them his latest results on condition that they were printed without delay; an arrangement that benefited both parties since many institutions abroad decided to subscribe to that hitherto little known journal. While Placzek was away I was the only one who knew enough Italian, and when each new copy of the *Ricerca* arrived I found myself the centre of a crowd demanding instant translation of Fermi's latest discoveries. And what an exciting time it was!

There was one group of elements which Fermi didn't have at

his disposal. They were the so-called rare earth elements, a dozen or so, which are chemically very similar and thus hard to separate; in those days only a few chemists had samples. Georg von Hevesy, a distinguished Hungarian chemist and a pioneer in the use of isotopes, was working at Bohr's institute; he had been given a complete set of oxides of those elements by the Auergesellschaft, the German chemical company that had first manufactured them mainly for use in the gas mantles which provided gas light before the electric light took over. So he asked one of his collaborators, a young woman called Hilde Levy, to expose his samples, one after another, to neutrons for a few hours and then test each sample with a Geiger counter.

My Geiger counter was available, complete with amplifiers and the necessary meter. It was very easy to place close to it a specimen which Miss Levy had irradiated, and take readings at regular intervals; from this we could deduce the half-life of any substance (half-life is the time in which the radioactivity of a sample falls to one half of its initial value) and also get some measure of the effectivity of the neutrons on that particular element. Most of them gave quite small effects, hardly measurable, but a few acquired quite respectable radioactivity. After we had gone through the whole series Hevesy wrote up a short account of those measurements for publication, but before sending it off he asked Hilde Levy – just to be on the safe side – to repeat the whole series.

One of the elements, dysprosium, had given a very small effect the first time we had tested it. In the second test it did so again; but as it was just time for lunch we left it while we went to eat. When we came back to the counter after lunch we immediately noticed that it was counting a bit faster; so we let it go on. Gradually more and more people came in and watched the amazing phenomenon: instead of getting weaker this preparation got more and more strongly radioactive as time went on!

We invented various wild theories before we realized what had happened. The dysprosium sample had become so strongly radioactive that it had almost choked the counter, making it count at about the same low rate as when there was no preparation present. It took several hours before the counter reached the full counting rate that it was capable of, rattling away at several hundred counts per minute; only then the rate began to decrease in the usual manner with a well-defined half-life the way it should. We all shuddered to think how easily we might have missed the discovery of the very strongest artificial radioactivity and hence a valuable detector for neutrons.

The study of atomic nuclei was pursued in other ways as well, by means of optical spectroscopy. Spectral lines which may look simple with a primitive spectroscope usually turn out to be clusters of lines: they have a fine structure or even what is known as a hyperfine structure, and this offers clues to the spin and the magnetism of atomic nuclei. That work was conducted by a Danish experimenter, Ebbe Rasmussen, with a group of collaborators. One of them was a gifted and cheerful German scientist with a quietly attractive young wife; his name was Hans Kopfermann, and so his wife was naturally known as 'die Kopferfrau'.

The team worked their way systematically through all the obtainable nuclear isotopes, and the analysis of each group of spectrograms took weeks. It had become a joke to ask the Kopferfrau beforehand what the spin of the current isotope would be, and though not a physicist herself she seemed to have an uncanny knack at guessing. In fact some of us said that asking her was a much simpler and just as reliable technique for determining the spins of isotopes. Kopfermann also played the violin rather well, and we often made music together.

Niels Bohr was fifty then and at the height of his powers, both mentally and physically. He was heavily built, but when he thundered up the stairs two steps at a time we young ones found it hard to keep up with him. He also beat us all at table tennis. There was a table in the small library, and the readers didn't seem to mind an occasional game. In his young days he had been an active

23. Niels Bohr (1885–1962), the Danish genius who (in 1913) first applied Planck's quantum ideas to the structure of atoms. A profound thinker who guided much of the growth of quantum theory.

football player, together with his younger brother Harald, himself a distinguished mathematician. Harald was the better of the two at football; for a time he played in the Danish national team. But whenever anybody, in his student days, commented on the brilliance of this young mathematician he would merely say 'I am nothing; you ought to meet my brother Niels.'

In appearance, Niels Bohr reminded you of a peasant, with hairy hands and a big heavy head with bushy eyebrows. I still remember his eyes which could hold you with all the power of the mind behind them; and then suddenly a smile would break over his face, turning it all into a joke. He was a yachtsman, he went skiing in the mountains of Norway, and I have seen him chop down trees, wielding a long-handled woodcutter's axe with the strength and precision of a professional. He usually went about on his bicycle. Once, returning from a long visit to Russia he had forgotten the combination of his bicycle lock. He still remembered a few of the levers which had to be pulled or pushed, and some of us remembered others, having watched him unlock his bicycle. Between us, putting our recollections together, we managed to reconstruct the combination after a hectic half hour so that Bohr could cycle home.

In the evening we often went to his house, the House of Honour given to him for life by the Danish Academy; a little palais built by the brewer Carl Jacobsen who had founded the Carlsberg Brewery in the last century. At dinner his wife Margrethe presided with unobtrusive efficiency and unfailing charm and kindness. After dinner, we would sit around Bohr, some of us on the floor at his feet, to watch him first fill his pipe and then to hear what he said. He had a soft voice with a Danish accent, and we were not always sure whether he was speaking English or German; he spoke both with equal ease and kept switching. Here, I felt, was Socrates come to life, tossing us challenges in his gentle way, lifting each argument to a higher plane, drawing wisdom out of us which we didn't know we had, and which of course we hadn't. Our conversation ranged from religion to genetics, from politics to modern art. I don't mean to say that Bohr was always right, but he was always thought-provoking and never trivial. How often did I cycle home through the streets of Copenhagen, intoxicated with the spirit of Platonic dialogue!

No other physicist of our time, except perhaps Einstein, has so strongly influenced our thinking in general, not just in physics. I have mentioned his model of the atom which brought him immediate fame in 1913; you know, the one with the electrons circling around the nucleus like miniature planets, confined to certain allowed orbits except when they jumped from one orbit to another in the process of absorbing or emitting radiation. That picture was so astonishing and unorthodox at the time that a number of physicists, and my old Hamburg boss Otto Stern among them, had sworn to give up physics if that nonsense was true (none of them did).

Bohr himself was very much aware of the crudeness of that model; it resembled the atom no more than a quick pencil sketch resembles a living human face. But he also knew how profoundly difficult it would be to get a better picture. Imagine you wanted to observe some shy animals that come out only at night. You might take one flashlight picture, but the flash would send them scuttling for safety. In the same way you could (in principle) take a flashlight picture of an atom; but once again, the flash would send the electrons flying. You might avoid upsetting the animals by using very weak light or infra-red light which they can't see, but with atoms there is no way out: to observe the position of an electron you would need at least one quantum of radiation, in fact an X-ray quantum, which would be enough to knock the electron clean out of the atom. Infra-red quanta are more gentle and might be used to measure the speed of an electron, much as the police use radar to check the speed of a car; but, just because of their long wavelength such infra-red quanta would tell us only roughly the position of the electron. If you try to measure the speed and position of an electron at the same time, both results will be beset by uncertainties, as indicated by Heisenberg's famous uncertainty relation.

Bohr saw this as a special case of what he called complementarity: the features of an atomic system form pairs – like position and speed – such that you can accurately observe each member of the pair but not both at once. Moreover Bohr suggested that the concept of complementarity may be of use outside atomic physics, that it may help us to a deeper understanding of the relation between matter and life, or body and mind, or justice and mercy. I can do no more than hint at those ideas, but I think they are

24. Niels Bohr with James Franck (middle) and Georg von Hevesy (in the light suit) who started and skilfully exploited the use of isotopes in biology.

important and will bear fruit in years to come. It is a bit as if reality was painted on both sides of a canvas so that you could only see one aspect of it clearly at any time.

Bohr never lost time with trivialities. Once when I visited him at his country cottage at Tisvilde, on the north coast of Zealand, I had brought a mathematical problem; one of those teasers which intellectual magazines are apt to publish, one which I had been unable to solve. Bohr first dismissed this as essentially trivial; but then to please (and perhaps teach) me he relented. He tackled the problem like a terrier shaking a rat. In a couple of minutes he had clarified its essential features and the best way to get the solution. After that he dropped it; the solution as such did not interest him.

But when it came to attacking a real problem, a serious problem of physics, he was marvellous to watch. I always felt that he moved with the skill of a spider in apparently empty space, judging accurately how much weight each slender thread of argument could bear. When he had explored the field his assurance grew and his speech became vigorous and full of vivid images. I remember an occasion when after a lengthy discussion on the fundamental problems of quantum theory a visitor said 'It makes me quite giddy to think about these problems.' Bohr immediately rounded on him and said 'But, but, but...if anybody says he can think about quantum theory *without* getting giddy it merely shows that he hasn't understood the first thing about it!' He never trusted a purely formal or mathematical argument. 'No, no' he would say 'You are not thinking; you are just being logical.'

Another great scientist I must talk about was James Franck. I had already mentioned him briefly as a life-long friend of Lise Meitner and as the scientist who, together with Gustav Hertz, got a Nobel Prize for first exciting specific states in atoms by bombarding them with electrons of controlled speed. He came to Copenhagen from Göttingen because he was a Jew. He could have stayed under the existing racial laws, which exempted men who had fought in the First World War, but it was unthinkable for him to serve under a regime that persecuted Jews. He had uncommonly fine features, luminous with kindness and obvious interest in your problems; he was the most immediately lovable man I have ever met.

In Copenhagen he was not too happy. For one thing he found the language difficult to pick up although he took lessons assiduously. There was one comic little episode when the postman rang his bell and asked something in Danish. Franck understood him and answered, but then realized from the blank face of the man that he was speaking English, not Danish. He tried a second time but, as he sadly reported 'It again came out in English!' In his work he was not too happy either. He had made his fame with the study of atoms and molecules and had an almost uncanny feel for what he might learn from an experiment in that field. But in Copenhagen he found us all busy with nuclear physics and, seeing that the neutron was the most promising tool for exploring atomic nuclei, he decided to plunge into neutron physics, and we all tried to help him. Once I joked that he would soon discover sharp energy levels in nuclei, like the ones he had found in atoms back in 1913, but he sadly shook his head and said 'No, no. That sort of thing happens only once. Nuclei are quite different.' Yet he was wrong: it was only a year later that sharp energy levels were found in nuclei, as close together as some of the levels which he had discovered in atoms. But of course the techniques required for their study were quite different.

When Franck realized that he had been following false trails in pursuing the neutron while others had been pushing ahead, he gave in: he accepted that at his age he had better not change his field. He went to Chicago and returned to the study of molecules, especially of the extraordinarily complicated mechanism by which chlorophyll (which gives plants their green colour) transforms carbon dioxide and water into organic molecules, using the energy of sun light. That was a problem right up his street; he unravelled the basic features of that mechanism – on which the life of plants depends – and opened the door to a large area of fruitful research.

Once I told him a crazy idea of mine whereby, with the help of a cloud chamber, one might look for an anti-neutron, in those days an object of pure speculation. Much to my surprise, a few days later Franck suggested the same experiment as we were sitting over our lunch. I said nothing; probably he had thought of it independently, and I wasn't sure if I had told him my idea. A few hours later he came to me and said 'Listen, Frisch, didn't *you* suggest that experiment? And if so, why didn't you pull me up, why didn't you say: But Franck I told you that – it was my idea!' Not many professors would so frankly expose their absent-mindedness – for that is what it was – to a much younger person.

In 1945 Franck's name became known to a much wider circle when he headed a group of Chicago scientists in urging the

American Government not to use the atomic bomb against a Japanese city but first to demonstrate its destructive power on a desert or an uninhabited island. There were good reasons why in the end the American Government rejected that proposal; but the 'Franck Report' will live as a reminder that James Franck was not only a great physicist but a resolute defender of human values.

Denmark 1934–1939: 2

Niels Bohr's fiftieth birthday – 7 October 1935 – was celebrated in style. It started in the morning with speeches given in the main lecture room, and though I got there early I found all the seats taken. So I went to the back, and leaned against a table; but soon the whole standing space had become so crowded that I pulled myself up on the table and sat cross-legged on it. That was fine at first, but to sit cross-legged on a wooden table gradually became pretty painful, and there was no way of getting off for the next two hours, with people standing packed and speeches going on without a pause.

In the evening there was a great banquet at the House of Honour. There were so many speeches, so many people talking about their recollections of Niels Bohr, of outings and discussions, of sailing trips and other adventures, that dinner advanced exceedingly slowly. After each dish there were several speeches, and though it went on till eleven the dinner was never finished. The next morning Mrs Bohr sent word to the laboratory that anybody who liked could come along and help eat up what was left.

The birthday was also used by George von Hevesy as an excuse to organize a grand gift from the Danish people to Niels Bohr. He managed to collect a hundred thousand Danish kroner, enough to buy about half a gramme of radium, which was presented to Bohr. With that radium it was possible to make a neutron source which would allow the physicists to do many interesting experiments; most of the neutrons, when the physicists didn't want them, were used by Hevesy himself to produce radioactive phosphorus for his biological experiments with radioactive tracer elements, a technique he had largely created and skilfully exploited.

In order to use radium as an efficient neutron source it has to be mixed with beryllium. We had a few chunks of metallic beryllium, a very hard, brittle silvery metal, which we ground up laboriously in porcelain mortars. Nowadays there are more convenient ways of doing that, and safer ones as well. Today we know that about one person in a thousand is highly allergic to beryllium and quite likely to die if he breathes small amounts of beryllium dust, and I shudder to think what might have happened in Copenhagen when I was given the job to get those chunks of beryllium, perhaps an ounce or so, ground up into a fine powder. I procured about a dozen pestles and mortars and got practically everybody in the laboratory to take turns at grinding beryllium. It must have been several hundred man hours before that hard metal was reduced to a really fine powder (which surprisingly looks quite black); fortunately nobody in the laboratory was allergic to it. Then the powder was mixed with a solution of a radium salt, and the mixture allowed to dry; I think Hevesy did that job himself.

Once the mixture had been divided into several small airtight capsules with long handles it usually led a peaceful life at the centre of a gallon-sized glass flask filled with carbon disulphide. In this way most of the neutrons had a good chance of striking one of the sulphur nuclei and producing the desired radioactive isotope of phosphorus which Hevesy needed for his experiments. Once every two weeks the flask was removed from its normal place at the bottom of a well in the institute, the radium source was taken out, and the liquid was distilled so as to concentrate the radioactive phosphorus. Carbon disulphide is highly volatile, so the distillation took only a few hours and a modest amount of heat, normally quite a harmless procedure.

But once, as I walked past the chemical laboratory, the girl who supervised the distillation came rushing out, coughing violently and screaming 'It's on fire!' Burning carbon disulphide produces choking fumes of sulphur dioxide, so I held my breath before I went to look. A big bluish flame hissed at the top of the broken neck of the flask; after some seconds I needed air again and had to leave the room in a hurry. The next person to appear was Hevesy who courageously clapped a wet towel round the top of the bottle in the hope of quenching the flame; but the only result was that the neck cracked off further down and the flame was more violent than before.

Then rescue appeared in the form of our workshop head, Mr Olsen. He came at a run, with one of his workshop boys carrying a cylinder of carbon dioxide with a rubber tube attached. This he directed against the flame, trying to blow it out, but it only waggled and hissed like an angry snake. But then he had an inspiration: he stuffed the rubber tube right into the neck of the flask, and the flame went out instantly. I still see Olsen standing there triumphantly holding down the rubber tube as if he dared the flask to burst into flame again.

Mr Olsen was a small, wizened chap but a very good mechanic, and, as you have just heard, a man of intelligence and initiative. One of his oddities was his pride in his trousers, which he called his thunderpants because whenever he got near a Geiger counter it started to rattle. Those trousers had a little radium solution spilt on them many years previously, and Olsen refused to part with them. They seem to have done him no harm.

There was another incident when that bottle of carbon disulphide became a menace. The girl was taking the freshly filled bottle down the spiral staircase to the bottom of the well where the radium was kept. The well had been dug years previously and carefully waterproofed so as to stay dry; its bottom, deep below ground level, provided a vibration-free location for a spectrograph, a delicate optical instrument which was no longer used there. Anyhow, at the bottom of the well the bottle slipped and broke. The disulphide immediately began to evaporate, and the girl fled up the iron staircase, pursued by those poisonous and highly inflammable vapours. Called to see what could be done, I cautiously went down the staircase, holding my breath. About half-way down my eyes began to sting; so I knew that the well was half full of an explosive mixture of air and the heavy carbon disulphide vapour. As I went up again I wondered whether my shoes would strike a spark from the staircase and cause an explosion which would blow me through the roof. Next we reflected how best to get those vapours out of the well, and the only safe way we could see was to use our biggest vacuum pump to suck out the vapour and blow it through a window into the open air where the wind would disperse it; but it would take a good many hours. My bedroom was at the top of the institute just about above that well, and I went to bed in a fatalistic mood. But the next morning I was still there and the danger was over.

Colloquia for the discussion of new experiments and new ideas were often called at very short notice; word passed around the institute and in a matter of minutes everybody was in the lecture room wondering who would talk and what about. They were very informal occasions, and I still remember once when the young Russian physicist Landau (whom I mentioned earlier as ticking off Einstein after one of his lectures) sat down on the lecture bench, tired from his talk, and then lay down flat on his back. In that position he continued arguing and gesticulating up at Niels Bohr, who was bending over him earnestly trying to convince him that he was wrong. Neither of the two appeared to be aware that this was a very unusual way of conducting a scientific discussion in front of an audience. After six years in the rather conventional atmosphere of Germany, it took me a while to get used to the informal habits at the Institute of Theoretical Physics in Copenhagen, where a man was judged purely by his ability to think clearly and straight. I might say that Bohr was far too polite ever to tell anybody that he was talking nonsense, but we soon learned what it meant when he said 'Very, very interesting'.

From time to time there was alarming news of some experimental result which appeared to contradict what we knew. Such a contradiction was an enemy immediately to be attacked, against which Niels Bohr turned the full power of his mind. Sometimes it turned out that the experiment had simply been wrong, and everybody was relieved. But on other occasions Niels Bohr would tell us one day with even greater delight that it was he who had made a mistake, that the inconsistency disappeared when one had found the right way to think about it. He never hesitated for a moment to admit that he had been in error. To him it merely meant that he now understood things better, and what could have made him happier?

Since we now had a neutron source we were able to repeat and extend some of the experiments which Fermi had done in Rome and which had puzzled us considerably. In particular there was his discovery that slow neutrons had so much more effect on certain nuclei than fast ones. In fact the whole phenomenon of neutron capture was hard to understand. According to what was then believed about nuclei, a neutron should pass clean through the nucleus, with only a small chance of being captured. Hans Bethe in the U.S.A. had tried to calculate that chance and I remember the colloquium in 1935 when some speaker reported on that paper. On that occasion Bohr kept interrupting, and I was was beginning to wonder, with some irritation, why he didn't let the speaker finish. Then, in the middle of a sentence, Bohr suddenly stopped and sat down, his face completely dead. We looked at him for several seconds, getting anxious. Had he been taken unwell? But then he suddenly got up and said with an apologetic smile, 'Now I understand it'.

The concept that had taken shape in those memorable few seconds has become known as the 'compound nucleus'. Bohr had realized that a neutron entering a nucleus would at once collide with one of the protons or neutrons inside, and those nucleons in turn would collide with others, so that the energy of the neutron would rapidly be dissipated and distributed among the other particles.

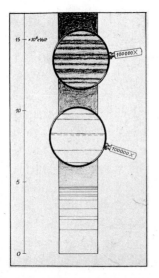

25. A drawing by the author, used by Niels Bohr to show how nuclear energy levels get denser and more diffuse at higher energies; above the dotted line a neutron can escape. Note that the lenses magnify 100000 times (I wish I knew how to make such lenses!).

One might say the nucleus would get hot. After that it must take a very long time – really a minute fraction of a second but long by nuclear standards – before by sheer chance one of the neutrons happened to get enough energy to escape. If during that time some energy got lost in the form of a gamma quantum – a high-energy photon – then there was simply no longer enough energy left, and the neutron was trapped for good. In other words, the nucleus can cool down by radiation, like a red-hot poker, and is then too cold to evaporate even one neutron. But that takes a relatively long time and the uncertainty principle allows the compound nucleus to possess sharply defined energy states; Fermi's experiments had already given some evidence pointing in that direction. James Franck, the first to prove that atoms had well-defined energy states, had not expected such states in nuclei; but Bohr's idea made their presence seem likely.

At the time I didn't understand much of that, but Placzek did. He persuaded me to join forces with him, and together we measured the absorption of slow neutrons in gold, cadmium, boron and various combinations of those elements. Placzek wanted quite thick layers of gold, and it was his inspiration to use several Nobel Prize medals which some of Bohr's German friends, when the Nazis came to power, had left in Copenhagen for safe keeping. It gave us great satisfaction that these otherwise useless slabs of gold should thus be employed for a good scientific purpose! By the way, the medals were once more in danger when Denmark was occupied by the Nazis, but were saved by Hevesy who simply dissolved them in acid and put them in a bottle; just one more of the many bottles containing evil-smelling chemicals. After the war they were turned back into metal, and Sweden struck them into medals once more.

Our joint work with Placzek showed that gold had a sharp 'resonance', a strong preference for capturing neutrons of a well-defined low energy, only a few electron-volts, many thousand times less than anyone had expected previously; and so had cadmium. But that was just what Bohr had come to expect from his idea of 'compound nuclei', and you can imagine how pleased he was about our result, and that he urged us to publish it quickly. But that was not so easy. Placzek and I had different ideas on how our results ought to be presented, and since he only woke up in the evening we had to do our writing at night; most of the final discussions were in the small hours when I was sleepy and stubborn. But after a couple of strenuous nights the text was finally agreed on, with Placzek still protesting faintly, and I took it to the nearby Post Office myself at four in the morning to prevent any further alterations. Placzek then suggested we should telephone Bohr, who had gone to dine with the King; he looked up the Royal Palace in the telephone directory, but desisted when I pointed out that Bohr had probably gone home.

My collaboration with Placzek was natural enough since he knew what he wanted to try and I had the experimental equipment, in particular a Geiger counter which was working reliably and could be used to measure the effects of neutrons on the various specimens he wished to have irradiated. The trouble was that we tended to keep different hours. I liked to get up about eight in the morning, whereas Placzek usually appeared at the laboratory around noon, yawning and rubbing his eyes. On the other hand, I preferred to go to bed before midnight whereas Placzek liked to work until three or four in the morning.

Hevesy suffered from insomnia and often came to the laboratory at night. One day he said with his deep Hungarian voice 'Frisch, I can see you around the laboratory at all possible hours. Do you never sleep?' So I explained the reasons for my staying up with Placzek, usually until three, trusting him to put the radium in its bottle and switch off the counters after I had gone to bed. Hevesy, with a faraway look on his long melancholy face, began to tell a story. 'In the village where I lived as a boy,' he said 'they once caught a young wild pig which had lost its mother. And they put the pig together with a young domestic pig, so that they should grow up in company; but that did not work. The wild pig was used to a nocturnal life and would rummage all night, keeping the tame pig awake; and the tame pig didn't let the wild pig sleep in daytime. So after a few weeks both pigs died.' Naturally, I related that sad story to Placzek who promptly accepted the title 'wild pig' as if it was an honour bestowed on him. He also tried to depreciate the fact that he was doing honest experimental work, and when I once pointed out to him that he was wearing a laboratory coat which made it clear to everybody that he had become an experimenter he hastily took it off. On the other hand, he wouldn't admit to being a theoretician either. He said 'I am an experimenter, or rather I used to be one. Then I stopped working, and since then people think I am a theoretician.' It was a sort of inverted boasting, because in fact he was quite a hard worker and produced important work on theoretical physics even though he kept unconventional hours.

Placzek was not a very tidy person (but who am I to throw stones?) and on one occasion was in despair because he had borrowed a manuscript from Niels Bohr and couldn't find it. He implored us all to look for the thing; he would be the tidiest man in the laboratory, he swore, if only the manuscript was recovered. Some of us insisted that this oath should be written up on the blackboard for all to see, and Placzek obliged, writing it down not only in English, but also in Cyrillic characters in Russian, and in Hebrew and Arabic for good measure. The manuscript turned up soon after. I cannot say that I noticed any difference in Placzek's tidiness although he stoutly maintained that he was true to his promise. Once I lent him my hat because heavy rain had started during the day. He didn't bring it back the next day or the day after, so I reminded him. He said he had mislaid the hat and offered to reimburse me. 'Well' I said, 'when I bought that hat it cost two pounds, but I have worn it for several years.' Placzek pulled out his wallet and handed me the equivalent of two pounds with the remark 'I'm no old-clothes merchant.'

In the meantime history took its course, and things happened which even I, a physicist in his proverbial ivory tower, could not disregard. Hitler occupied Austria, and suddenly I was no longer an Austrian; there was no longer such a thing. I reported to the police and was told that I had to apply for a German passport; much as I disliked becoming one of Hitler's subjects, there was no way around it. I was afraid that the Danes would refuse to renew my permit, as previously they had done every year; but the only change I recall was that it now had to be done twice a year. I don't know if Niels Bohr spoke on behalf of the refugees (of whom I was by no means the only one) who were working in his institute.

Actually, shortly before Austria was taken over, I had booked a holiday trip together with a lanky American physicist, Jackson Laslett, who had spent several months in Denmark quietly learning Danish without ever using it in the laboratory, and spending most of his time with his feet up on a table, his chair tilted back, and on his lap a drawing board on which he drew unperturbably one component after another of a cyclotron we were building. He had been 'lent' to us by Ernest O. Lawrence in whose laboratory in California the first cyclotron – a big atom-splitting machine – had been built. Anyhow, we had planned a little trip through Sweden and Norway, where the language was not too different from Danish.

Austria was taken just as we set out. Fortunately by then my Danish was good enough to persuade the Swedish immigration officer into letting me in, despite my Austrian passport which was now of no value as a travel document. But he advised me to speak to the Norwegian Consulate in Gothenburg before attempting to cross over into Norway. That cost us two days because we arrived in Gothenburg on a Saturday. Still, we met some friends and enjoyed our weekend there, and on Monday the Norwegian official was cooperative and let us continue our journey, seeing we both had complete round-trip tickets, and willing to believe me that I was indeed going back to Denmark. Our train took us right across the central mountain range to the Norwegian town of Bergen, far up in the north, and as the train slowly wound its way up to the watershed we felt that we really wanted to spend at least a few hours among those lovely sun-drenched hills. With sudden decision we jumped out of the train with our small amount of luggage; the train had left when we asked how soon there would be the next one. 'In twenty-four hours', was the answer!

So there we were stuck for a whole day, in Finse, right on top of the watershed; the mild, sunny weather had given way to a snowstorm; as it was June we had not even overcoats, and the hotel, half a mile from the railway station, was closed because it was not the skiing season. We were lucky enough to persuade a janitor to open up and let us have a room with two beds in which we spent most of those twenty-four hours because it was too cold to go out. How glad we were to be back in the train the next day!

In Bergen – having got there three days late – we looked at our tickets and decided sadly that most of the beauties they offered us – such as the famous Hardanger Fjord – would have to be by-passed because Laslett felt he had to get back to his work. So we had to spend some more money by flying from Bergen to Stavanger, another fjord but not nearly so beautiful. It was my very first flight, disappointingly like a ride in a noisy country bus, full of Norwegian peasants, but with a fine view of snow-covered mountains on our left. From Stavanger we continued by bus (as our tickets said) to the southern tip of Norway and arrived pretty exhausted and half seasick from a trip of several hours with innumerable hairpin bends. There we almost immediately embarked on a boat that took us across the Kattegat to Denmark. It was my stormiest journey ever; we were both sick all through the night. Even when we were again on terra firma the land appeared to sway, and the waves that hit the cliff with a nose like cannon shots rose in a spray a hundred feet high. We relaxed at the house of Professor Buch Andersen, a Danish colleague of mine, but Laslett insisted on taking the train back to Copenhagen the next day while I was lazy and stayed a day or two more with that delightful family, rolling on the lawn in mock fights with their two lively children.

Hitler had occupied the Rhineland, and even I began to realize that the whole balance of Europe was tottering; that we could not expect that precarious peace to last much longer, and that sooner or later Hitler would occupy Denmark, and I would be back in the frying pan. Placzek said coolly 'Why should Hitler occupy Denmark? He can just telephone, can't he?' It was a cruel joke, but not far from the truth; when the Nazis eventually invaded Denmark the military resistance was called off after a few hours. Early in 1939 the threat was there for all to see.

From then on, whenever an English visitor came I put out feelers if a job could be found in England where I would be out of the real danger of being sent to a concentration or extermination camp and where I might even have a chance to help with the fight against that menace to civilization. In the meantime I continued my work in Copenhagen, but in a half-hearted way. I felt that my fruitful and happy time there was drawing to a close and that whatever I started would not be completed. But then came the great surprise: the discovery of uranium fission.

Energy from nuclei

Did I say that all nuclei had weights that were multiples of that of a hydrogen nucleus? That is not quite true; most of them are about 1% lighter than that, and therei :ies the secret of nuclear (often called 'atomic') energy. When protons come together to form heavier nuclei their joint mass becomes less by an amount m, and a lot of energy E is set free, following Einstein's formulae $E = mc^2$. The factor c^2 (speed of light multiplied by itself) is very large, so a minute amount of mass corresponds to a lot of energy; for instance the mass of a paper-clip is equivalent to the entire energy a small town uses during a day.

Energy is measured in a variety of units: kWh (kilowatt-hours) on your electricity meter, Btu (British thermal units) for the gas man, and so on. Those are man-size units, much too large for a single nucleus. For them the common unit is the MeV (a million electron-volt, but usually we say 'an emmeevee'). It is the energy of motion which an electron (or a proton) acquires when it is accelerated by a voltage of a million volts. An alpha particle has typically 5 to 10 MeV; to keep a watch going needs several million times as much energy every second.

Einstein's formula was put to the test in the 1930s by measuring the energy of the particles (e.g. protons) set free in 'atom splitting'. The collision of two nuclei caused the nucleons to be rearranged so as to form two new nuclei; when those were both of a kind found in nature it was possible to compare the masses of the nuclei before and after the collision and check the mass difference against the energy set free. That was done with mass spectrographs which were soon made so precise that Einstein's formula could be checked to within a fraction of an MeV; it was always found correct when the nuclei formed by the reaction were stable and hence available for mass spectroscopy. When unstable nuclei were formed one had to take into account the energy of the particles they subsequently sent out in transforming themselves into stable nuclei again. Soon there was a network of literally thousands of measurements, cross-checking each other, and the masses of several hundred isotopes were accurately known.

What do those masses tell us? Well, for one thing, they tell us why the sun keeps shining. If you could dive into the huge white-hot ball of not-quite-pure hydrogen which we call the sun you would find rapidly rising pressure and temperature until near the centre the temperature is around ten million degrees Centigrade. At such heat the hydrogen nuclei move so fast (about 500 km/sec) that they occasionally collide despite their mutual electric repulsion. There are traces of other elements, which complicate what happens; Hans Bethe, whom I later met in Los Alamos, was the first to work out a possible mechanism for this process in detail. To cut the story short, the main outcome is simply that helium nuclei are formed, one from four hydrogen nuclei (two of which are changed from protons into neutrons), and each hydrogen nucleus gives up 7 MeV in that process. In this 'nuclear fire' about a million times more energy is produced than in ordinary (chemical) fire, for instance when hydrogen burns by combining with oxygen. Even so the amount of hydrogen the sun has to burn to keep shining is stupendous: about ten billion tons every second! But the sun is big: in the four billion years since the Earth became solid the sun has used up only a fraction of its hydrogen.

If you go on to build up heavier nuclei you still liberate energy, but much less, and stars that run out of hydrogen become unstable.

26. Hans Bethe,
German-born master of all
aspects of theoretical physics,
particularly as applied to
nuclei (Nobel Prize 1967).
(Sketch by the author, about
1944.)

That raises fascinating questions regarding the nature of novae, supernovae, pulsars and so on. But here I'm getting on thin ice (or into hot water?), so let us return to solid ground.

Here we have some simple clues. Light nuclei contain as many neutrons as protons. The reason is a variant of Pauli's housing rule: two protons, spinning oppositely, can inhabit one quantum state, together with two neutrons behaving the same way. The first complete family of that kind is indeed the helium nucleus, rare on Earth but exceedingly common in the sun and the stars. But then why do heavier nuclei contain relatively more neutrons? Why is the ratio of neutrons to protons about 1.2:1 for copper, 1.4:1 for iodine and 1.6:1 for uranium? Because protons are bad club members: they are electrically charged and hence repel each other, and it makes a heavy nucleus more stable if some of them are turned into neutrons even though, as a result, they may have to move into

111

higher quantum states. Nuclei with too few or too many protons adjust the ratio after a while by sending out an electron or a positron, as I mentioned earlier.

But in the heaviest nuclei, even when the ratio of neutrons to protons is at its optimum, the protons are still under pressure from their mutual repulsion. Then why don't they just get pushed out? In fact what is holding nuclei together? The protons repel each other, and the neutrons – being electrically neutral – cannot be held by electric forces. Gravity? Many million times too weak.

Today we know that any two nucleons attract each other very strongly, but only when they are very close together. We have no special name for that attraction; we call it simply 'the nuclear force'. It is more like a kind of stickiness, and we even think we know something about the nature of the glue. It acts only between nucleons in the same nucleus, except for a brief moment when two nuclei collide.

But the heavy nuclei have a trick to unload some of their quarrelsome protons. Two protons can combine with two neutrons and emigrate as a family; the 28 MeV which are gained (as in the process that keeps the sun shining!) serve to pay for the exit visa, as it were. In classical mechanics such a process would be impossible; like mountaineers trying to climb out of a crater on an insufficient supply of food, they would find that their energy gives out before they reach the rim and overcome the pull of the other nucleons.

Classical physics is adamant about that, but the laws of quantum mechanics are more flexible. They allow our subatomic mountaineer to 'tunnel' through the crater wall, as some physicists like to put it. Or you may imagine that two protons and two neutrons use Heisenberg's uncertainty principle to borrow some energy, to be repaid after they have left the nucleus and become a helium nucleus, a newborn alpha particle, rapidly driven away by the electric repulsion of the remaining nucleus, sliding down the outer crater wall as it were. But such a loan is granted only after

27. Lise Meitner, lecturing.

uncounted billions of applications; in other words the chance of an alpha particle to escape in any given split second is minute and depends of course on the kind of nucleus. That chance was calculated from Schrödinger's wave equation, by Edward Condon (U.S.A.) with Ronald Gurney (U.K.), and also by the Russian, George Gamov, in 1926.

Until 1938 nobody dreamt that there was yet another way for a heavy nucleus to react to the mutual repulsion of its many protons, namely by dividing itself into two roughly equal halves. It was mere chance that I became involved in the discovery of that 'nuclear fission', which for the first time showed a way to make huge numbers of nuclei give up their hidden energy; the way to the atom bomb and to atomic power.

The occupation of Austria in March 1938 changed my aunt, the physicist Lise Meitner – technically – from an Austrian into a German. She had acquired fame by many years' work in Germany, but now had to fear dismissal as a descendant of a Jewish family. Moreover, there was a rumour that scientists might not be allowed to leave Germany; so she was persuaded – or perhaps stampeded – into leaving at very short notice, assisted by friends in Holland,

28. Fritz Strassmann, the German chemist, who with Otto Hahn discovered the fission of uranium in 1938.

and in the autumn she accepted an invitation to work in Stockholm, at the Nobel Institute led by Manne Siegbahn. I had always kept the habit of celebrating Christmas with her in Berlin; this time she was invited to spend Christmas with Swedish friends in the small town of Kungälv (near Gothenburg), and she asked me to join her there. That was the most momentous visit of my whole life.

Let me first explain that Lise Meitner had been working in Berlin with the chemist Otto Hahn for about thirty years, and during the last three years they had been bombarding uranium with neutrons and studying the radioactive substances that were formed. Fermi, who had first done that, thought he had made 'transuranic' elements – that is, elements beyond uranium (the heaviest element then known to the chemists), and Hahn the chemist was delighted to have a lot of new elements to study. But Lise Meitner saw how difficult it was to account for the large number of different substances formed, and things got even more complicated when some were found (in Paris) that were apparently lighter than uranium. Just before Lise Meitner left Germany, Hahn had confirmed that this was so, and that three of those substances behaved chemically like radium. It was hard to see how radium – four places below uranium – could be formed by the impact of a neutron, and Lise Meitner wrote to Hahn, imploring him not to publish that incomprehensible result until he was completely sure of it. Accordingly Hahn, together with his collaborator, the chemist Fritz Strassmann, decided to carry out thorough tests in order to make quite sure that those substances were indeed of the same chemical nature as radium.

When I came out of my hotel room after my first night in Kungälv I found Lise Meitner studying a letter from Hahn and obviously worried by it. I wanted to tell her of a new experiment I was planning, but she wouldn't listen; I had to read that letter. Its content was indeed so startling that I was at first inclined to be sceptical. Hahn and Strassmann had found that those three substances were not radium, chemically speaking; indeed they had found it impossible to separate them from the barium which, routinely, they had added in order to facilitate the chemical separations. They had come to the conclusion, reluctantly and with hesitation, that they were isotopes of barium.

Was it just a mistake? No, said Lise Meitner; Hahn was too good a chemist for that. But how could barium be formed from uranium? No larger fragments than protons or helium nuclei (alpha particles) had ever been chipped away from nuclei, and to chip off a large number not nearly enough energy was available. Nor was it possible that the uranium nucleus could have been cleaved right across. A nucleus was not like a brittle solid that can be cleaved or broken; George Gamov had suggested early on, and Bohr had given good arguments that a nucleus was much more like a liquid drop. Perhaps a drop could divide itself into two smaller drops in a more gradual manner, by first becoming elongated, then constricted, and finally being torn rather than broken in two? We knew that there were strong forces that would resist such a process, just as the surface tension of an ordinary liquid drop tends to resist its division into two smaller ones. But the nuclei differed from ordinary drops in one important way: they were electrically charged, and that was known to counteract the surface tension.

At that point we both sat down on a tree trunk (all that discussion had taken place while we walked through the wood in the snow, I with my skis on, Lise Meitner making good her claim that she could walk just as fast without), and started to calculate on scraps of paper. The charge of a uranium nucleus, we found, was indeed large enough to overcome the effect of the surface tension almost completely; so the uranium nucleus might indeed resemble a very wobbly, unstable drop, ready to divide itself at the slightest provocation, such as the impact of a single neutron.

But there was another problem. After separation, the two drops would be driven apart by their mutual electric repulsion and would acquire high speed and hence a very large energy, about 200 MeV in all; where could that energy come from? Fortunately Lise Meitner remembered the empirical formula for computing the masses of nuclei and worked out that the two nuclei formed by the division of a uranium nucleus together would be lighter than the original uranium nucleus by about one-fifth the mass of a proton. Now whenever mass disappears energy is created, according to Einstein's formula $E = mc^2$, and one-fifth of a proton mass was just equivalent to 200 MeV. So here was the source for that energy; it all fitted!

A couple of days later I travelled back to Copenhagen in considerable excitement. I was keen to submit our speculations – it wasn't really more at the time – to Bohr, who was just about to leave for the U.S.A. He had only a few minutes for me; but I had hardly begun to tell him when he smote his forehead with his hand and exclaimed: 'Oh what idiots we all have been! Oh but this is wonderful! This is just as it must be! Have you and Lise Meitner written a paper about it?' Not yet, I said, but we would at once;

and Bohr promised not to talk about it before the paper was out. Then he went off to catch his boat.

The paper was composed by several long-distance telephone calls, Lise Meitner having returned to Stockholm in the meantime. I asked an American biologist who was working with Hevesy what they call the process by which single cells divide in two; 'fission', he said, so I used the term 'nuclear fission' in that paper. Placzek was sceptical; couldn't I do some experiments to show the existence of those fast-moving fragments of the uranium nucleus? Oddly enough that thought hadn't occurred to me, but now I quickly set to work, and the experiment (which was really very easy) was done in two days, and a short note about it was sent off to *Nature* together with the other note I had composed over the telephone with Lise Meitner. This time – with no Blackett to speed things up – about five weeks passed before *Nature* printed those notes.

In the meantime the paper by Hahn and Strassmann arrived in the U.S.A., and several teams did within hours the same experiment which I had done on Placzek's challenge. A few days later Bohr heard about my own experiments, not from me (I wanted to get more results before wasting money on a transatlantic telegram!) but from his son Hans to whom I had casually talked about my work. Bohr responded with a barrage of telegrams, asking for details and proposing further experiments, and he worked hard to convince journalists that the decisive experiment had been done by Frisch in Copenhagen before the Americans. That was probably the source of the story – reprinted several times – that I was Bohr's son-in-law (although he never had a daughter, and I was then unmarried). I can see how it happened: a journalist asks: 'How do you know of this, Dr Bohr?' Bohr: 'My son wrote to me', Journalist mutters: 'His son, but name is Frisch; must be son-in-law'.

During this turmoil in the U.S.A. we were quietly continuing our work in Copenhagen. Lise Meitner felt that probably most of the radioactive substances which had been thought to lie beyond uranium – those 'transuranic' substances which Hahn thought they had discovered – were also fission products; a month or two later she came to Copenhagen and we proved that point by using a technique of 'radioactive recoil' which she had been the first to use, about thirty years previously. Yet transuranic elements were also formed; that was proved in California by Ed McMillan, with techniques much more sensitive than those available to Hahn and Meitner.

In all this excitement we had missed the most important point: the chain reaction. It was Christian Møller, a Danish colleague, who first suggested to me that the fission fragments (the two freshly formed nuclei) might contain enough surplus energy each to eject a neutron or two; each of these might cause another fission and generate more neutrons. By such a 'chain reaction' the neutrons would multiply in uranium like rabbits in a meadow! My immediate answer was that in that case no uranium ore deposits could exist: they would have blown up long ago by the explosive multiplication of neutrons in them. But I quickly saw that my argument was too naive; ores contained lots of other elements which might swallow up the neutrons; and the seams were perhaps thin, and then most of the neutrons would escape. So, from Møller's remark the exciting vision arose that by assembling enough pure uranium (with appropriate care!) one might start a controlled chain reaction and liberate nuclear energy on a scale that really mattered. Many others independently had the same thought, as I soon found out. Of course the spectre of a bomb – an uncontrolled chain reaction – was there as well; but for a while anyhow, it looked as though it need not frighten us. That complacency was based on an argument by Bohr, which was subtle but appeared quite sound.

In a paper on the theory of fission that he wrote in the U.S.A. with John Wheeler, Bohr concluded that most of the neutrons emitted by the fission fragments would be too slow to cause fission of the chief isotope, uranium-238. Yet slow neutrons did cause fission; this he attributed to the rare isotope uranium-235. If he was right the only chance of getting a chain reaction with natural uranium was to arrange for the neutrons to be slowed down, whereby their effect on uranium-235 is increased. But in that manner one could not get a violent explosion; slow neutrons take their time, and even if the conditions for rapid neutron multiplication were created this would at best (or at worst!) cause the assembly to heat up and disperse itself, with only a minute fraction of its nuclear energy liberated.

All this was quite correct, and the development of nuclear reactors followed on the whole the lines which Bohr foresaw. What he did not foresee was the fanatical ingenuity of the allied physicists and engineers, driven by the fear that Hitler might develop the decisive weapon before they did. I was in England when the war broke out, and in Los Alamos when I saw Bohr again. By that time it was clear that there were even two ways for getting an effective nuclear explosion: either through the separation of the highly fissile isotope uranium-235 or by using the new element plutonium, formed in a nuclear reactor. But I am again getting ahead of my story.

Birmingham 1939–1940

My departure from Denmark was not at all dramatic. I had received a letter from Mark Oliphant, head of the department of physics at the University of Birmingham, inviting me to come over during the summer vacation and discuss what could be done for me; I had expressed my worry to him when he visited Denmark some months previously. So I packed two small suitcases and travelled by ship and train, just like any tourist. Oliphant had impressed me by his aura of confidence and calmness; in his presence you felt that nothing could go wrong and everything necessary would be done without fuss. Let me relate an example. Once, in the tea room, when we were all standing around and discussing physics, Oliphant put down his tea cup and said 'We really must have a blackboard in here, you can't argue without one', and he turned to one of his assistants to locate a blackboard that wasn't in use. The assistant came back a few minutes later and said he had found one, but it was too large to get up the narrow stairway. 'Never mind' said Oliphant 'we can get it through the window.' He told somebody else to find a block and tackle; they went up to the roof, fixed it, hoisted the blackboard out of one window and in by the window of the tea room, and within half an hour we had a blackboard. Elsewhere that would have required days of planning and quite likely some paper work.

But to come back to my first weeks in Birmingham. It was a

29. Mark Oliphant, the Rutherford pupil who in Birmingham directed part of the development of radar and later built the first high-energy accelerator in England, then returned to his native Australia. Knighted in 1957.

pleasant, hot summer, and I seem to have spent much of my time lying in the sun; there was very little else to do. Oliphant had offered me a job as auxiliary lecturer; but it was still holiday time, with no lectures. The international situation was uncertain, and anything might happen any day; it would be rash to go back to Denmark. I think we all imagined scenes out of H. G. Wells' *The Shape of Things to Come*: a fleet of aeroplanes dropping thousands of bombs, buildings toppling, millions fleeing, all within a day and probably without a declaration of war.

It didn't come like that, but the situation grew more and more ominous. Hitler had been presented with an ultimatum, asking him either to withdraw his troops from Poland or to consider himself at war with Britain; we all sat round the radio and there was a great feeling of tense sobriety when we were told that the deadline had passed and that the war was on. But for quite a long time there were no toppling buildings, no screaming bombs. Hitler knew quite well that we could not stop his advance into Poland, and he went ahead with that business before turning to the West. All that is a matter of history, and I won't write about it; others have done that much better than I could.

For a while I stayed with a landlady near the university; the usual bed-and-breakfast arrangement, having my meals mostly elsewhere. The winter was unusually severe, and we had several weeks when the temperature did not rise above freezing and even fell below zero Fahrenheit, that is below minus eighteen degrees Centigrade: very unusual for Britain though I had experienced even colder weather in Berlin. A lot of snow had fallen and every morning I walked through the small front garden between walls of snow which went up to my hips.

The job that Oliphant had allotted to me was very informal. He was giving lectures to beginners and was doing it in a casual and almost improvised manner, relying on his great skill in making things plausible without much preparation. But he did realize that for some students it might well be heavy going, and so at the end of the first lecture he had told them 'Now, for those of you who find they are rather at a loss and would like to ask questions, Dr Frisch will be in lecture room number three and will try to help you.'

Those sessions were lively and much to my taste. Usually I found a couple of dozen students, when I came in, and to keep things informal I just sat on the lecture bench, dangling my legs and saying 'Well now, who wants to start shooting questions?' There usually was a short silence, and then they came with a rush. I remember one occasion when the first question came from a girl who said 'Tell us about viscosity'. 'What about viscosity?' I answered. 'All about viscosity!' she said. That was a tall order, but I did my best to help them understand the notes they had taken in Oliphant's lecture.

In those days there were plenty of exciting problems related to nuclear fission. But we were not well equipped for that sort of work; Birmingham didn't have a cyclotron then. Moreover, Oliphant had decided to concentrate on the development of what we then called radiolocation, later to be known as radar; that is, the technique of observing the direction and the distance of an object such as an enemy aeroplane by the reflection of very short radio waves from it. That work was so secret that we foreigners (there were a few of us) were not allowed to take part or even to know what was going on. Much of that secrecy was a bit of a charade. I remember at tea time Oliphant casually approaching my colleague, Rudolf Peierls, who was a first-rate but German-born mathematical physicist, and saying 'If you were faced with the problem of solving Maxwell's equations for a cavity with conducting walls in the shape of a hemisphere, could you cope with it?' And Peierls would say 'Well, it's an interesting problem; I'll give it some thought', and a day or two later he would come back and say 'I have a solution for that problem you gave me.' Now Peierls knew that this was connected with the generation of very short electric waves, such as were needed for radar, and Oliphant knew that Peierls knew, and I think Peierls knew that Oliphant knew that he knew. But neither of them let on; they both pretended that this was purely an academic mathematical problem that had occurred to Oliphant out of the blue and with which he had challenged Peierls, perhaps to test his ability as a mathematical physicist. I had no such mathematical gifts and couldn't take part in the work even to that limited extent.

All the same I would, of course, have liked to do some research and, naturally I tried to find a suitable problem related to uranium fission. Now one of the urgent questions was whether Niels Bohr was right with his surprising proposal that the observed fission of uranium by slow neutrons consisted in splitting not the common uranium isotope uranium-238 but the isotope uranium-235, more than a hundred times rarer. The only sure way to test that was to prepare a sample of uranium in which the proportions of the two isotopes were changed. So I began to read up about possible methods of separating isotopes and came across one which looked attractively simple; I felt that I might be able to build such a device with just a little help from the strained resources of a department that was working all out on developing radar.

The method had been invented by a German, Klaus Clusius, and in its simplest form merely required a long tube standing upright with an electrically heated wire along its axis. This tube was to be filled with a gaseous compound of the element whose isotopes one wanted to separate, and the theory of Clusius (which had been confirmed by experiments) indicated that material enriched in the lighter isotope would accumulate near the top of the tube while the heavier isotope would tend to go to the bottom. All simplicity itself.

Becoming familiar with the theory was quite hard mathematical work for several weeks; it showed me, as in fact Clusius knew, that for optimum efficiency it was better not to use a thin wire but a thick rod or tube which was heated by some means such as having an electric heater element down the centre. Such a glass tube was made for me by the glass blower, and I was proposing to test the experiment in the first place by just leaving air in the tube. Nitrogen, which is lighter, should accumulate near the top and oxygen near the bottom, if it worked. (Of course there are simpler ways of separating oxygen and nitrogen; this was just a way of testing the tube.)

But the work went rather slowly; the glass blower had constant demands from the radar boys and couldn't spare much time for me. Still, Oliphant made sure I wasn't neglected altogether. There was also the problem of space; all the laboratories had been taken over for the radar development. But there was a small unused lecture room where I was allowed to set up my equipment.

While this was going on I was unexpectedly invited to write an article for the Chemical Society for their Annual Report on Progress; I was to report on advances in nuclear physics in so far as they were of interest to chemists. I managed to write that article in my bed-sitter where in daytime, with the gas fire going all day, the temperature rose to $42°$ Fahrenheit (about $6°$ Centigrade) while at night the water froze in the tumbler at my bedside. What I did was to pull a club chair up close to the gas fire, wear my winter coat and put the typewriter in my lap so as to be protected from all sides; the radiation from the gas fire stimulated the blood supply to my brain, and the article was completed on time.

Of course it contained a section on nuclear fission, what little was known about it at the time; it mentioned the exciting possibility of a chain reaction, but also reported Niels Bohr's reassuring argument that an explosion of really violent character would not be produced even if it became possible to assemble a mass of uranium which would sustain a growing chain reaction. Bohr argued that the neutrons which are sent out from uranium in the fission process would be too slow to cause splitting of the prevalent isotope uranium-238 but would simply get swallowed up by the isotope. Only a small fraction of the neutrons would be caught by the light isotope uranium-235 which then would undergo fission and produce a few more neutrons, not enough for a chain reaction.

The efficiency of neutrons in causing fission can be greatly increased by slowing them down, and experiments were going on in France and the U.S.A. to see whether with the help of those slow neutrons a self-supporting chain reaction might be achieved; a question which was still wide open at that time. But even if that was possible, Niels Bohr pointed out, it takes a significant time to slow down neutrons and therefore such a chain reaction could only grow at a moderate rate. It might be a way of making a source of energy, but not an effective bomb. At the worst the material would heat up rapidly, melt, and some of it evaporate; the reaction would then stop because the neutrons would escape, having liberated no more than a microscopic fraction of the energy that was available in the mass of uranium. The result would be no worse than setting fire to a similar quantity of old-fashioned gunpowder.

I had felt very reassured, like most of us, when Niels Bohr had put forward that argument, and naturally I related it in my report, which was duly sent off and printed.

People have often asked me whether that was a piece of deliberate camouflage. I can assure you it wasn't. I really believed what I wrote; that an atomic bomb was impossible. But after writing that report I wondered – assuming that my Clusius separation tube worked well – if one could use a number of such tubes to produce enough uranium-235 to make a truly explosive chain reaction possible, not dependent on slow neutrons. How much of the isotope would be needed? I used a formula derived by the French theoretician Francis Perrin and refined by Peierls to get an estimate. Of course I didn't know how strongly fission neutrons would react with uranium-235, but a plausible estimate gave me a figure for the required amount of uranium-235. To my amazement it was very much smaller than I had expected; it was not a matter of tons, but something like a pound or two.

Of course I discussed that result with Peierls at once. I had worked out the possible efficiency of my separation system with the help of Clusius's formula, and we came to the conclusion that with something like a hundred thousand similar separation tubes one might produce a pound of reasonably pure uranium-235 in a modest time, measured in weeks. At that point we stared at each other and realized that an atomic bomb might after all be possible.

I have often been asked why I didn't abandon the project there and then, saying nothing to anybody. Why start on a project which, if it was successful, would end with the production of a weapon of unparalleled violence, a weapon of mass destruction such as the world had never seen? The answer is very simple. We were at war, and the idea was reasonably obvious; very probably some German scientists had had the same idea and were working on it. A German scientist, Gustav Hertz (the one who pretended to drink a beaker of pure alcohol), had been one of the first to separate isotopes (of neon, not uranium) in significant quantities, and the possibility was well known to the physics community. So Peierls and I went to talk it over with Oliphant. He told us to write it all down and send our report to Henry Tizard, who was advising the government on scientific problems concerned with warfare. That report was sent off within a couple of weeks and was decisive in getting the British Government to take the atomic bomb seriously; it had already been discussed, chiefly by George Thomson, whose father had discovered the electron in 1897.

The scientists who were gradually drawn into that work were mainly not British scientists but refugees, most of whom had not yet been naturalized. The reason why so many of Hitler's refugees took part in the work on the atom bomb is widely believed to have been a desire for vengeance; but that was not so. The main reason was that most British physicists who were willing to do war work (and not all of them were) had already started on other things, chiefly on the development of radar which appeared to be the most promising means to defend the country from enemy aircraft.

In the meantime I had been unaware of another danger hanging over me: the danger of being sent to an internment camp. After all, that was the normal thing to do with an enemy alien like myelf – send him to an internment camp unless there was a good reason why he should not be interned. So when one day I received a summons from the police I went with some misgivings, which were not dispelled when I listened to the questions: 'Do you have any dependent relatives?' 'Are you about to sit for an examination?' 'Are you about to get a degree which would give you a better chance of work?' Even though I was sent home after the interview I felt that all those questions really added up to the simple question 'Is there any reason not to intern that chap?' Mrs Peierls felt sure I would be interned and made me buy some shirts of sea-island cotton which could be washed by a bachelor like me; but fortunately I never had to wash them. I went to see my friend and colleague, Philip Moon (deputizing for Oliphant who had gone away for a few days) and asked him if he couldn't persuade the police that I was engaged on important war work. He said he would try, and obviously succeeded because I never heard from the police again.

My attempt to separate uranium isotopes was in its early stages, and even if it worked it would take some time before I had enough uranium-235 to measure directly how strongly it was affected by neutrons. So I considered an indirect method, namely using a source of fairly slow neutrons which – if Bohr was right – would not cause fission in the common isotope uranium-238; the fission rate observed would then be entirely due to uranium-235. Such a neutron source can be obtained by surrounding – not mixing! – radium with beryllium which is thus affected only by the penetrating gamma rays. Radium is very expensive, but from it one can extract the gas 'radon'; a good source of the gamma rays we wanted. Radon decays with a half-life of four days but is formed anew at the same rate in the radium from which it has been extracted. At my request Oliphant arranged for me to get some radon from a hospital in Manchester; the radium had been removed to safety, deep below ground in the Blue John Cavern in Derbyshire, a well-known tourist attraction in peace time.

So one day I went by train to Manchester and was taken from the hospital by car to the cave. Down I went over slippery ladders and through narrow, muddy passages to a slightly larger cavity where, incongruously, there was a laboratory table with a lot of glassware on it, bulbs and tubes and stopcocks, rather like the equipment I had used in Hamburg. That was the plant for 'milking' the radium, for extracting the radon and compressing it into a small glass capillary, no longer than half an inch. At Oliphant's request the radium had not been milked for a whole week so that a large amount of radon had accumulated. Less than an hour later, when the local technician had done the work for me, I walked out with my little suitcase containing a heavy block of lead at the centre of which was this tiny capsule full of radon, equivalent in radiation to about three-quarters of a gramme of radium. Any safety officer today would shudder at the thought that I walked out with that thing, protected by only a couple of inches of lead, and that I travelled within a few feet of that radiation source first by car and then by train. Today that would be considered an unacceptable radiation hazard both to myself and to other people in the compartment.

When I got the radon back to Birmingham I instantly started measuring its effect (surrounded by beryllium) on an ionization chamber containing a gramme or so of uranium, a chamber I had designed specially for that purpose. I continued the measurements without interruption for thirty-six hours, taking naps of half an hour on a bed with an alarm clock at my side. Then the measurements were stopped because the source was running down (and so was I).

That ionization chamber led to quite an important discovery. The electronic equipment for it had been constructed by Ernest Titterton, who has since become Professor and head of the physics department at Canberra in Australia as well as being knighted, but who at that time was a young student, very bright and active, whom Oliphant had detailed to help me with the work. I kept complaining that the chamber from time to time produced a pulse which looked just like a fission pulse but surely couldn't be; there was no source of neutrons. Or was there? We went so far as to search the laboratory to see if by any chance a small source had been left in a drawer. We tested the electronics and tried all kinds of improvements. Nothing made any difference, and in the end I had to admit that uranium was occasionally suffering fission spontaneously. Because of the war and the secrecy surrounding our work Titterton could not publish that important discovery; it was made about the same time by the two Russian physicists G. N. Flerov and K. A. Petrzhak who are generally quoted as the discoverers of spontaneous fission.

My marathon thirty-six hour measurement had given a smaller fission rate than expected; I had over-estimated the effect of neutrons on uranium-235 and consequently under-estimated the amount that would be needed for a bomb. Fortunately Peierls had in the meantime discovered a simple way of reducing that amount, namely by surrounding the uranium with a material that turns some of the neutrons back so that they don't escape unused. As a result we were really not much worse off than originally estimated.

I don't remember much of my everyday life in Birmingham these days. At first I stayed in lodgings, but later Peierls and his Russian-born wife Genia invited me to stay with them; they had just moved into a larger house. That was very pleasant, comfortable and entertaining. Peierls's sister-in-law used to toast her bread at breakfast at the gas fire behind her, using a toasting fork; calls of 'it's smoking' she ignored, and only a scream of 'it's on fire!' made her withdraw the fork, watch her bread burn for a while, then blow it out and eat the cinder.

Genia ran her house with cheerful intelligence, a ringing Manchester voice and a Russian's sovereign disregard of the definite article. She taught me to shave every day and to dry dishes as fast as she washed them, a skill that has come in useful many times since. Even while I was still in lodgings I often spent an evening with the Peierlses and then had to walk home in the blackout, a distance of nearly two miles. If there was a moon or even a reasonably clear sky it was not difficult; indeed I found it peaceful to be alone with the stars. But there were nights when on shutting the front door I felt that I must have gone blind, when I could not even tell the sky from the ground. Even so it is amazing how adaptable one is. Usually I walked home those two miles at normal walking speed without any incident, except that once on a particularly dark night I fell over a bench that I had forgotten was standing on the pavement. Many people had taken to wearing luminous cards in their hatbands; they glowed faintly and prevented people from colliding with one another. But lampposts didn't wear them. Motor cars were visible, but hardly more; their head lamps were hooded in a way that allowed only a glimmer of light to escape, just enough to warn a pedestrian they were coming, and to show the driver a few feet of road ahead.

The report which Peierls and I had sent to (Sir) Henry Tizard on Oliphant's advice had triggered the formation of a committee, with (Sir) George Thomson as chairman, which was given the code name 'Maud Committee'. The reason for that name was a telegram which had arrived from Niels Bohr, ending with the mysterious words 'AND TELL MAUD RAY KENT'. We were all convinced that this was a code, possibly an anagram, warning us of something or other. We tried to arrange the letters in different ways and came out with mis-spelt solutions like 'Radium taken', presumably by the Nazis, and 'U and D may react', meant to point out that one could get a chain reaction by using uranium in combination with heavy water, a compound of oxygen and the heavy hydrogen isotope called deuterium, abbreviated D. The mystery was not cleared up until after the war when we learned that Maud Ray used to be a governess in Bohr's house and lived in Kent.

At first those two enemy aliens, Peierls and I, were not allowed to take part in the deliberations of the Maud Committee, but that was obviously inefficient; not only had our report started the whole thing, but we had also thought about many of the additional problems that would arise; so it was reasonable that we should be consulted. In the end a sub-committee was formed, of which we were members and which was set up to discuss technical questions, leaving the higher political decisions outside our ken, at least officially. One of these Maud meetings (they took place in London) sticks in my mind. For the first time an American colleague, visiting London because of his work on guided missiles, had been invited to take part; we wanted to get the Americans interested in our work. He was Charles Lauritsen, a Dane who had left his country as a youngster and then without any formal training worked his way up to the position of full Professor at Caltech, the California Institute of Technology in Pasadena. After the meeting I walked with him along Piccadilly to his hotel, explaining all sorts of details which I felt he might not have taken in during the meeting. If you think that was risky I disagree; I can think of no better place than the milling crowds of Piccadilly for transmitting secret information – in Danish.

The first time I visited Liverpool I went with Peierls, on the invitation of (Sir) James Chadwick (discoverer of the neutron). He was in touch with I.C.I. (Imperial Chemical Industries) and had asked them about the possibility of producing uranium hexafluoride, the only gaseous compound of uranium known to be stable enough to be put into a Clusius tube. At the physics department of the university we were shown to Chadwick's study; he came in after a short while, sat down at his desk and started to scrutinize us, turning his head from side to side like a bird. It was a bit disconcerting, but we waited patiently. After half a minute he suddenly said 'how much hex do you want?'. That was his way; no formalities, straight down to brass tacks.

On returning I pointed out to Oliphant that my experiments would be hampered in his institute where radar must take precedence, whereas the Liverpool institute had not started any war work and moreover had a cyclotron which would serve as a powerful source of neutrons whose energy we could control. Liverpool was a 'prohibited area', not open to enemy aliens; but Chadwick was willing to have me and able to make the necessary arrangements with the authorities.

Liverpool 1940–1943

It was in Liverpool that for the first time I heard the wailing of air-raid sirens. To begin with people followed instructions and went quickly into shelters or houses; but when nothing happened except that within a few minutes the steady note of the all clear was sounded they began to ignore the sirens, at least in daytime. At night it was different: the sirens were soon followed by the popping of anti-aircraft guns and sometimes the clatter of falling shrapnel; in our boarding house we were all pretty frightened and went into the basement, as instructed. But that again wore off after a while, and I remember often spending part of the night in the lounge, playing pontoon (vingt-et-un) with a couple of officers who for some reason had nothing better to do. At other times I practised the piano; some of the lodgers said the noise was worse than the guns.

Then the bombardment began in earnest, and when one emerged in the morning one found that well-known buildings were now empty shells, a few walls still precariously standing; and soon a mobile crane would come hobbling along, swinging a large steel ball against those dangerous walls until they collapsed into a cloud of dust and rubble. Some of the architecture students in my boarding house helped with the demolition work; as one of them remarked, it taught them a lot about architecture. All the same, life went on and we got used to it; we knew that the real target

30. Collaboration between the U.K. and U.S.A., based on mutual respect: to the left, James Chadwick (1891–1974) who discovered the neutron in 1932; head of the British atomic-energy mission to the U.S.A. in 1943. Nobel Prize 1935, knighted in 1945. To the right, Leslie Groves, the U.S. general who was in overall charge of the atom bomb project.

of the air-raids were the docks, several miles away from where we lived, and that therefore the chance of us being hit was remote.

But one night a violent bombardment of the town itself began. We could hear the bombs whistling as they fell and we huddled under the staircase, which was generally regarded as the safest place. There was one bomb which kept whistling until we thought it must be our turn; a painfully loud explosion was followed by the tinkle of falling glass as most of our windows had been blown in. At that point our landlady decided to quit and didn't even bother to collect outstanding rents; she just disappeared. I packed my suitcases and proceeded on foot – no tram, too many bomb craters – to the house of a friend of mine, the mathematical physicist Maurice Pryce, who lived six miles from the centre of Liverpool with his wife Gritly, daughter of Max Born. When after my two

hours' walk I put down my cases in their hall the Pryces agreed
that I could stay but they were going to leave soon; they had a new
baby and were trying to go to a quieter place.

I found I could make myself useful by talking to the baby when
he started crying in the early morning. No doubt he would have
preferred milk, but was willing to accept a bit of chat and tickling
and eventually went to sleep again when he saw that no milk was
forthcoming. But I had been there only a couple of nights when
I was woken by the words 'Frisch, your house is on fire!' I had
left some things at the boarding house, and the voice was that of
Tom Chalmers, a medical physicist and also air-raid warden. He
regarded air-raids as a huge joke and high adventure and had driven
through the raid to the Pryces' house in order to fetch me so that
I could collect my belongings. By the time we drove into town it
was lit up by the scattered fires, but otherwise quiet, and the 'all
clear' came soon after. Actually it wasn't my house that was on
fire but the church next door; the firemen had managed to save
the adjacent houses. One of the men told me that for a time he
had stood on his ladder, feeling a fool because no water came from
his hose; the pressure had fallen due to overload, with so many fires
to be fought at once. I picked up my typewriter (there was not much
else to be collected) and Chalmers drove me back to my quiet
suburbs.

That whole trip, by the way, was illegal, because I was under
curfew like all aliens and not allowed to be out in the open after
darkness. That order rather hampered my work because in order
to get home before curfew I often had to stop a measurement that
was going well and could have been completed in an hour or two.
Moreover I wasn't allowed to use a bicycle. Chadwick had full
sympathy with my difficulties when I explained them to him and
he persuaded the Chief Constable of Liverpool to waive those
restrictions for me so that thereafter I was allowed to stay out and
cycle home at any time of night.

When some days later the Pryce family left, they allowed me to stay in the house and even bequeathed me their cleaning woman. So for quite a while I lived luxuriously in a pleasant residential district where I could sleep in peace. My slumber was disturbed only once when I was woken by voices in the garden and came fully awake at the words 'We'll just have to break the window'. Hastily jumping out of bed and sticking my head out I shouted 'What's the matter?' The two air-raid wardens on the brightly illuminated lawn looked up. 'Ah, someone is in' one of them called back; 'what about those lights?' Only then did I tumble to the enormity of my negligence: I had left the lights on in the living-room and forgotten to draw the blackout curtains! Fortunately it was a quiet night with no enemy aircraft in the vicinity; the wardens accepted my agitated apologies, and nothing more was said.

Things were more complicated when one summer night I was cycling home during curfew hours, having by then obtained permission to do so. I was stopped by a constable and was getting ready to explain my peculiar position; but all he was concerned about was that I was riding without lights. I assured him that it was still light enough to see, but he said that it was past lighting-up time and proceeded to take down my name and address. All the time I expected that at any moment he would spot my foreign name and accent and ask me 'What the devil are you doing here on a bicycle *and* after curfew?' But I was speaking my best English, and he probably took me for one of those Londoners who to a Liverpudlian anyhow sound like foreigners. So eventually I was allowed to go home, having turned on my lamps.

A couple of weeks later I received not one summons but two: one referring to my front lamp and one to my rear lamp! I rather fancied myself standing in Court and pleading my case, but unfortunately the summons was for a date right in the middle of a short holiday which I had planned to spend with the Peierlses in Cornwall. So I went to the police and asked what I should do and whether the date could be changed. No, the date could not be changed: but I was advised to write a personal note to the magistrate, pleading guilty and assuring him I wouldn't do it again. I was fined ten shillings, five for each lamp.

There were worse offences which I committed quite unwittingly. An old friend from Vienna, Dr Herz, a heart specialist who had been nicknamed the Herz-Herz (because Herz means heart in German) had settled in Wales, not very far from Liverpool, in the town of Mold. As I was very fond of him, I went to visit him every few Sundays. I took a bus to Chester and usually spent a little time walking around that picturesque town with its old cloisters and its lovely cathedral, before catching another bus to Mold where I spent a pleasant afternoon with that wise old man. He had one oddity, perhaps from past times when he had many appointments to keep; he often wanted to know the exact time, and he carried two watches to make quite sure.

Coming back was sometimes more difficult; by the time I was on the bus from Chester to Liverpool the air-raid sirens had gone, and some bus drivers simply refused to drive into the bombardment. They stopped somewhere outside Liverpool and said 'Everybody get off, the bus doesn't go any further'. Then the only way to get home was to try and cadge a lift, and many a time have I entered Liverpool on top of an army lorry; an odd vehicle for an enemy alien to ride! Moreover I discovered after the war that Chester was like Liverpool, a prohibited area which enemy aliens were not allowed to enter, and a different one from Liverpool; so I ought to have had permission from two Chief Constables each time I undertook the journey! But I was never picked up; I didn't even know that I was breaking the law.

In the laboratory it took me a while to find where everything was, and Chadwick had thoughtfully allotted one of his students, John Holt, to help me. In those days I was an energetic, fast-moving fellow, and with the student usually trailing behind me I found that we had become nicknamed 'Frisch and Chips'. I hasten to add that 'Chips' has since become a full Professor in the University of Liverpool. Together we built a small Clusius separator and found that it gave no measurable separation; other scientists confirmed that uranium hexafluoride is one of the gases for which the Clusius method does not work. Fortunately Peierls had reckoned with that possibility and had thought up another method, together with Franz Simon, which used diffusion through porous membranes; it was more complex but sure to work, and did. (Gustav Hertz used a similar method to separate neon isotopes.)

Then there was Gerry Pickavance, who was busy building a mass spectrometer, a young man with a very pleasant personality with whom I became good friends. He had a flair for solving technical problems and organizing work, and he later became head of the Rutherford Laboratory, in charge of building a big atom splitter. It was a great distress to his many friends when in 1970, still at the height of his powers, he was struck by a haemorrhage and became more or less incapable of speech. It seemed particularly awful to see that sort of thing happen to a lively, likeable and articulate person. Incidentally, as a motorist he combined a high average speed with a singular freedom from accidents; occasionally he said that he would have loved to be a racing driver.

At first I was a little afraid of John Moore, the head technician, because he tended to speak his mind rather sharply. Once I seem to have said so to someone who then told him. Next thing, I had John Moore asking me angrily what libellous remarks I had been making; I got angry too and we had quite a row, after which we became the best of friends. After the war he was put in charge of the construction of the big Liverpool cyclotron, which replaced the small one they had when I was there.

That I had a private life outside the laboratory I largely owed to Joseph Rotblat. He was a Polish physicist who had been on study leave in England when the war broke out. Of course he lost contact with his family when his country was overrun and divided between the Germans and the Russians, and his wife who had tried to flee had died or been killed. One might expect such a man to be embittered, but there was nothing of that to be noticed. He was a kind, outgoing person, always looking after others, always trying

31. Joseph Rotblat, Polish-born experimental physicist who later specialized on the medical effects of nuclear radiations. For many years secretary of the Pugwash Conferences which brought together scientists from East and West to seek ways of banishing the menace of nuclear war. (Caricature by the author.)

to help people. He later became Secretary General of the Pugwash Conferences and has done as much for peace as anybody I have ever encountered.

I remember one occasion when he had arranged for me to play the piano to a group of Polish soldiers stationed in Liverpool. When I got there I found a classroom with something like a hundred people crammed into it, and an old upright piano with half a dozen keys not working. One trial made me give up the idea of playing any delicate tune, which would be ruined if one note didn't come through; I would have to play music consisting largely of octaves and making a lot of noise. Indeed I had the cheek to play Chopin's Grande Polonaise, really far too difficult for me. But to those Poles, I still think, it meant a lot. For them the Grande Polonaise is almost like a National Anthem, and to hear it played in a foreign land was a rousing experience; there was a storm of applause when I had finished my very ragged performance.

Joe also got me in touch with English people, and Evan Gill

deserves particular mention. Among his several brothers were two artists, one physician and a bishop; he told me he was the only inconspicuous one, just a clerk in a big flour mill. But he was a cultured man with many interests, and one of his brothers was Eric Gill, the famous sculptor and designer of type faces, whose alphabets were used by *The Times* as well as by the London Underground, and who had a great influence on the design of clear and well-proportioned characters. I never met any of Evan's brothers, but he would from time to time give me little prints (wood engravings) by his brother Eric, which I still have and cherish. He was also working on a complete bibliography both of Eric's work and of the island of New Guinea which one of his brothers had explored. He lent me a book about it. It must be an incredible sort of territory, so full of crags and cracks that it might take a day to cover half a mile; to explore and police such a country must have been a job for heroes. After every few pages I had to shut the book and get my breath back.

In Evan Gill's house I soon became a regular visitor where we made music together; he played the violin, handicapped by an injury suffered in the First World War, but still well enough to make the music enjoyable. Often we were joined by a lively young clergyman, Gordon Robinson, who played the piano quite well, and I borrowed a spare fiddle to join in. My own violin had been left in Denmark and I didn't see it again until after the war.

The first time I was at Evan Gill's I enquired on leaving how I could best get back to town; they lived a couple of miles from the centre. Mabel, his wife, told me where the bus stop was – only a few minutes from their house – but when I asked her what bus to take she frowned and said she didn't really know. The way she did it was to go to the bus stop and get on when she recognised the driver! I thought that was a very English way of picking the right bus.

Another occasion that taught me something about the English way of doing things was when a large parachute bomb landed in the courtyard of the physics department. I had spent the evening constructing a very delicate instrument when at the last moment, by an inadvertent movement, I broke the glass fibre and went home ill-tempered in the knowledge that I would have to start all over again the next day. But during the night Maurice Pryce, who was fire watching on top of the clock tower of the university, saw a parachute descend, carrying one of the dreaded 'land mines' (a ton or so of high explosive), and saw it land near the foot of his tower. He squatted behind the parapet and waited for the explosion; when nothing happened for several seconds he cautiously raised his head to peer over. At that moment he saw a blinding flash and ducked. A hail of fragments whizzed past his head, and practically all the windows and some of the doors of the physics building were blown in; fortunately the tower stood.

When I arrived for work the next morning I saw the damage. My first reaction was 'Oh well, that glass fibre would have broken anyhow; let's go home until the building is repaired.' Then I saw that there were some people around, doing things such as nailing large sheets of cardboard over the broken windows; had I better stay and see if I could help? Who, I asked, was in charge of the work? It took me a while to grasp that nobody was in charge; this was spontaneous cooperation, people just doing what was needed. I asked one of them whether there was anything I could do, and he said 'Yes, go over to the engineering department and see if you can scrounge some more hammers!' This I did and then spent the rest of the day helping to nail cardboard over the windows and over the holes that had been torn in the roof. Fortunately it was a fine day; at the end of it the building was usable, and the next morning we were back at work. I remember thinking at the time that if that sort of thing had happened in Germany everybody would have gone home and waited until the appropriate authority had detailed a brigade of repair men to come and repair the building properly. But perhaps I am doing the Germans an injustice; maybe in conditions like that they also learned to improvise.

One surprising feature of the way our work was conducted was the lack of any outward sign of secrecy; no guards, no safes. The institute was apparently devoted to various kinds of innocent research and of course very largely to the teaching of physics to students. Rotblat even included fission in his lectures and mentioned the possibility of a chain reaction, but in such a casual way that nobody would have thought that it might lead to an important development in weaponry, let alone that we were actually working on that. I think it caused some worry among the more security-minded people; but it certainly worked. Any spy could have walked into the laboratory and asked questions, but of course we wouldn't have talked about our work to strangers, and moreover it didn't occur to any spy that we might deserve his attention. It has been confirmed again and again that nobody outside the institute had any inkling of the kind of work that went on.

Moreover, Chadwick often gave me a chance to discuss things with colleagues elsewhere, putting no great trust in the bogus security which relies on compartmentalizing knowledge, on letting every scientist know only what he needs to know. That kind of security leads to inefficiency. In fact I was from time to time sent off to London or Cambridge. In Cambridge a separate effort had started, largely under the influence of two French scientists who had come over from Paris on the very last boat that succeeded in leaving France after the German occupation, bringing along most of the world's heavy water. This allowed them to continue their experiments in Cambridge and to show that a combination of uranium and heavy water could indeed produce a chain reaction.

They were not Frenchmen by birth. Hans von Halban was Austrian by origin and a skilful organizer, who immediately took the leading role because he spoke English and knew how to impress people as a man of the world. His colleague was Russian-born Lew Kowarski, a bear of a man, with whom I had to talk French at first. But he learned English with amazing speed and soon acquired a command of its finer points which a great many Englishmen would envy.

32. Lew Kowarski, Russian-born member of the French team who first showed the possibility of nuclear chain reactions; built the first nuclear reactor in both Canada and France.

When I came to Cambridge, which often meant twelve hours' journey in the disruption caused by air-raids, I usually spent a night or two at the Kowarskis'. They had rented an apartment a little way from the centre, with a pleasant large lawn. Once, when I came, Lew proudly told me they had had a bomb the night before. I was duly impressed, and he offered to show me where it fell; taking me out in the middle of the unblemished lawn he pointed to one spot and said 'Here it was; no' he said, pointing to a place about two feet to the right, 'here'. It had been an incendiary which had not gone off and had been removed, leaving no trace. However at another time a bomb next door did a lot of damage to their apartment.

Cambridge did suffer some damage from bombs because many government departments had been moved there, and several people were killed, but there was hardly any damage in Oxford where I worked for a couple of months at the Clarendon Laboratory, under the guidance of the German-born scientist, Franz Simon – later knighted as Sir Francis Simon. He was a distinguished physical chemist who had been rescued from the anti-semitic regime of Hitler by his old colleague, Professor Lindemann (later Lord Cherwell). He combined a dry sense of humour with a profound understanding both of physics and chemistry; it was he and Peierls who worked out a way of separating uranium isotopes (by diffusion through porous metal membranes) which was actually used for the early bombs and is still in use to make enriched uranium, the fuel in many of the atomic power stations that were built after the war.

I was a little amused to find that in Oxford all the fire precautions and the organization of air-raid wardens' rosters and so on were treated more solemnly than in Liverpool, where fire and rubble had taught people to rely more on their ability to improvise; but this is by the way. The purpose of my visit was to develop a method for analysing small samples of uranium which had been enriched by an unknown amount that had to be determined. It took longer than I had thought, and the method did not work very well. When I came back to Liverpool I had thought of a better method which I proposed to Chadwick with some diffidence because it seemed an ambitious idea requiring several dozen radio valves. Today that may seem a very modest project; at the time I was quite surprised when Chadwick immediately agreed. But after looking at the problem of building the apparatus we decided to hand the job over to a branch of I.C.I. (Imperial Chemical Industries), which was within easy range – about an hour by train from Liverpool – and there I went from time to time to check progress and to make the machine work when it was ready.

The machine worked very well, and those little excursions were a great treat to me. When I got off the train I was collected by a large car and taken to a luxury canteen where some officers who were cooperating with I.C.I. introduced me to such strange beverages as Drambuie; luxuries which I had never encountered before. But it was really the success of the work which I enjoyed most, and the instrument, which depended on having a dozen amplifiers working in parallel, each one counting only alpha particles of a particular energy, turned out to be a very versatile research tool – the pulse height analyser or kicksorter – which was later improved out of all recognition by other people and is now a standard instrument in nuclear physics laboratories. Even the primitive instrument which was built under my supervision at I.C.I. turned out to be very useful for comparing the various methods of uranium separation.

The worst bombardment of Liverpool (as I learned after the war) was an attempt by the Germans to save their battleship *Bismarck*. They knew that the communications centre of the Royal Navy was located in the basement of the Post Office building in Liverpool, and they succeeded in burning that Post Office down to the ground; but the communications centre kept functioning, and the *Bismarck* was hunted down and sunk. I remember the next day, half in a daze, walking through the streets of Liverpool and watching the fires and the overloaded fire services, trying in vain to cope. While I was sitting on a park bench I saw a little flame lick at the door of the public library opposite me. As I watched, the whole building was gradually engulfed by the flames; there was no fire engine available to deal with that. I remember feeling glad that I had not returned some books, which therefore were safe in my room and could be given back later; a small consolation when the library itself was destroyed.

My stay in Oxford provided a peaceful interlude and a number of new friendships. I met the Hungarian physicist Nicholas Kurti, a bachelor with a great love and knowledge of fine cooking which he kept up even after his marriage some years later. The other person who sticks out in my memory was Shull Arms, a big lanky American from Idaho with the physique of a lumberjack; his huge hands were able to perform the most delicate operations, controlled by a very subtle brain.

Back in Liverpool I found new lodgings with a friendly landlady, Mrs Gibbs, a good way from the town centre, and was settling down to making improved measurements of the interaction of neutrons with uranium when one day Chadwick came to me and in his usual direct manner asked 'How would you like to work in America?' I said 'I would like that very much'. 'But then you would have to become a British citizen.' 'I would like that even more', I said. After that, things began to happen with bewildering rapidity. Within a few days a policeman appeared and started to take down personal data as well as the names of people who knew me and could vouch for me, explaining that he had been instructed to start naturalization proceedings. He added in an oddly confiding manner, 'You must be a pretty big shot; I have been told to get everything done within a week!' And indeed it was only about a week later that I got instructions to pack all my necessary belongings into one suitcase and to come to London by the night train, to a government office in Old Queen Street.

There a tall lady secretary acted like a maître de ballet, sending off each scientist on a new errand as soon as he came back from the last one. I was sent to a magistrate, who took my Oath of Allegiance to His Majesty the King; in return he gave me a document which stated that I was now a British citizen. Back at the office I was told to go and obtain release from military service; otherwise I was a deserter! Then I was sent to pick up a passport, ready for me, and with that I went to the American Embassy where somebody was waiting to stamp my visa into the new passport. I also visited a 'censor' who went quickly through the contents of my suitcase and – to my surprise – started to read one of the very few Danish books which I had packed in order not to get out of practice. But I had not packed the grammar from which for several months I had been studying Russian, under the influence of the battle for Stalingrad and our admiration for the grand resistance of our brave allies. I had almost reached the point when I could tell a simple story in my own words, but thought it unwise to bring a Russian grammar into an American army camp. By now I have forgotten what little I had learnt.

With the formalities all completed in one day I returned to Liverpool again on the night train, and the next morning we all boarded the *Andes*, a luxury liner equipped to bring American troops and almost empty on the westward journey. I had forgotten my 'ticket' but the head of our group, Wallace Akers, waved me through. There was a dormitory with eight berths for each of the few dozen scientists, some with families; being single I had a whole dormitory to myself. We stayed in harbour for another two days (presumably to confuse enemy spies), shamefacedly eating unheard-of delicacies such as grapefruit and fried eggs, surrounded by grimy docks and wartime austerity.

The *Andes* did not sail in convoy but relied on her speed and a zigzag course to evade submarine attack; the journey took twice as long as usual, and the climate changed from tropical to arctic and back. Everyday we had life-belt drill and our anti-aircraft guns were fired to make sure they worked. There was said to be a fine grand piano, but it was boarded up to protect it from the troops. All I had to play on was an old upright, strapped to one of the pillars in the saloon to prevent it from falling over when the weather got rough. My piano chair, which had little wheels, would occasionally start rolling away in the middle of a passage, and I had to grab the keyboard and pull myself back. Some of us got seasick, but otherwise the journey was uneventful, and the ship arrived safely, with perhaps the greatest single cargo of scientific brain-power ever to cross the ocean.

Los Alamos 1943–1945: I

We landed in Newport News, and there is a story that the immigration officer became suspicious of my Austrian accent, my German-sounding name and my brand new passport, issued and visaed on the day I became a British citizen; and that there was some consultation before I was allowed to go on the north-bound train with the others. But I remember nothing of that. The only thing that sticks in my memory is a glimpse of Richmond, Virginia, where we had to change trains and I wandered out into the streets. There I was greeted by a completely incredible spectacle: fruit stalls with pyramids of oranges, illuminated by bright acetylene flares! After England's blackout, and not having seen an orange for a couple of years, that sight was enough to send me into hysterical laughter.

In Washington we were told that it would be several days until General Leslie Groves could come to brief us before we were taken to our place of work. My friends had nicknamed me 'the man with an aunt in every port'; so I took a train to New York and visited my mother's youngest sister Frida Frischauer who taught mathematics at Adelphi College and was very surprised to see me turning up like that. Of course I couldn't tell her where I was going or why; she realized it must be secret work and didn't ask. It was my first sight of New York and a very pleasant interlude, though I only stayed one day.

148

When Groves finally arrived he told us where we were going – not all of us to the same place – and that I was going to Los Alamos; he spoke at length about security precautions which we would have to accept to ensure that nothing about our work got known to the enemy. Apart from that I remember very little either of the long train journey to Los Alamos in New Mexico, or of our reception there.

It was only a year previously that the U.S. Army had started to develop the Los Alamos Ranch School, a private school for boys, into a laboratory town which by the end of the war had grown to a population of some eight thousand; several hundred scientists with their families and a large number of supporting personnel. The bomb materials – uranium-235 and plutonium – were made elsewhere. The purpose of Los Alamos was to bring together the mathematicians, physicists, chemists and engineers who would determine the amount and the most effective arrangement of material needed; they would also design and test all the many devices required for its assembly into an explosive unit in which the chain reaction, triggered by a single neutron, would grow with lightning speed. The chosen site was near the centre of the United States, in New Mexico, among steep-walled canyons, accessible only by one rutted road; as isolated a place as one could wish for the most secret military establishment in the U.S.A.

I was given a room in one of the original school buildings, a magnificent blockhouse which was traditionally built from big treetrunks and generally known as the Big House. (It no longer exists.) Here, on the whole, were the quarters for bachelor scientists. Married ones were given individual houses, which were going up at a great rate. I admired the way they were built, quite primitive, just wooden planks nailed to a framework of timber on top of a foundation of a few concrete blocks, but the workmen used hand-held buzz saws to cut those planks on the spot with incredible speed, and once the concrete had set the house would be built in a couple of days.

33. The Big House (no longer in existence) in Los Alamos (N.M.) where many of the bachelor scientists lived, engaged in atom bomb research.

I remember watching one very large laboratory building going up. Standing there with one of my earliest friends in Los Alamos, Philip Morrison, who walked with a stick but had a very nimble mind and was very entertaining, we stared at the building for a while, and I said 'How are they going to heat that?', because it was early December and rapidly getting cold, and I couldn't see where the heating plant would go. 'Oh' he said 'It's such a large building, they'll probably just set fire to a corner of it, it won't make any difference.' Philip Morrison is now Professor of Astrophysics at the M.I.T. (Massachusetts Institute of Technology, Cambridge, Mass.). I also met Robert Oppenheimer early on, the distinguished theoretician who was scientific director of the site and used to greet newcomers with the words 'Welcome to Los Alamos, and who the devil are you?' His slight figure with the broad-brimmed porkpie hat was unmistakable. I found out later that it was he who had picked the site, at the edge of a large extinct volcano and some twenty miles or so by a winding, rutted road from the nearest town,

34. Robert Oppenheimer (1904–1967), U.S. theoretical physicist who became director of the Los Alamos laboratory in 1942.

Santa Fe, and thus in a very isolated situation at about seven thousand feet above sea level. Moreover Oppie had recruited not only the chemists, physicists and engineers that the project required but also a painter, a philosopher and a few other unlikely characters; he felt that a civilized community would be incomplete without them. The scientists that had come included some of the very cream of American universities, and I had the pleasant notion that if I struck out on any evening in an arbitrary direction and knocked on the first door that I saw I would find interesting people inside, engaged in making music or in stimulating conversation. I have certainly never seen a small town with such a variety of intelligent and cultured people.

The landscape too enthralled me from the beginning. During the winter I was usually at my breakfast table in time to watch the sun rise. There in front of the window was the rugged chain of the Rocky Mountains, a dark silhouette about thirty miles away. The sky above them grew lighter and lighter; the lightness began to contract to one particular point; and then suddenly, with blinding intensity, the first little segment of the sun. Within two minutes the breakfast room was filled with brilliant sunshine, every morning; all through the winter there was hardly ever a cloud to be seen, except for the occasional snowstorm which supplied what we needed for skiing. In the evening we could see the mountain chain turn red as the sun sank below the horizon, a lovely spectacle which had given the mountains the local name of el Sangre de Cristo. It was fascinating country, unlike anything I had ever seen.

During the summer there were the red flowers of the cacti, each a fiendishly spiny bush which grew to a height of six feet or more and then died off in the autumn leaving only the skeleton, each branch a network of wavy wooden ropes forming a very stiff girder. They grew tiny yellow fruits, and even they were covered with thousands of needle-sharp prickles. Behind us rose the heavily wooded rim of the Valle Grande, an extinct crater of over a hundred square kilometres, said to have been bought by a Spanish family for $20, a century ago.

The autumn colours of America are famous and some of the mountain-sides, where aspens grew, turned an unbelievable buttercup yellow in the autumn, such as I have never seen anywhere else, in brilliant contrast to the dark green of the pine woods. Below us in the valley was the Rio Grande in its early course, a quiet trickle of water during much of the year (and of course frozen in winter) but a torrent of tomato soup in the spring when it was fed by the melting snow of the Rocky Mountains and carried millions of tons of red soil. The ground in the valley had been cut up by ages of erosion into table mountains, some of those mesas almost unclimbable, with steep rocky walls like the Lost World which Conan Doyle so vividly described. I did manage to climb some of them but found no prehistoric monsters on top. Los Alamos itself was on one of the larger mesas, between two deep canyons.

On those climbing adventures I was usually with a Swiss friend whom I had first met in Cambridge, by name of Egon Bretscher; he combined physics and chemistry, had worked with Rutherford and was probably the first man to foresee the use of plutonium. In Cambridge he was regarded as a hypochondriac ('I don't know what is wrong with me today; I feel fine!' he is reported to have said once) but he was a keen mountain climber, and the sight of an unscaleable wall was an irresistable challenge to him. On one hot summer day, climbing with nothing but our swimming pants and our boots on, we managed to get up one of the more inaccessible mesas; but later, as the sun was sinking, we couldn't find the place where we had reached the top. We saw no other possible route of descent and got scared of facing a night up there, when even in summer it could get very cold under the clear sky. It was getting dark when we decided to strike out in the opposite direction from the one we came, and we were lucky; the slope there was gentle and we got back to the valley by the light of the moon. We were getting pretty cold by then, and there was another hour's brisk walk round the mesa before we got back to my car.

The site was dominated by the water tower; a cylindrical wooden container like a huge barrel, without any architectural pretensions. In spite of that we had a water shortage one winter. The story was that a soldier had been sent down the pipeline to check the condition of a valve, but had misunderstood his orders and closed it. By the time he had come back and reported, only to be sent down again to reopen it, the water in the pipe had frozen, and all attempts to unfreeze it in the severe cold – we often had sub-zero temperatures in winter – were in vain. The water-level in the tower began to fall; instructions were issued to economize, and after a while the water supply was turned off at certain hours. These included the hours when a man likes to shave, and beards began to sprout all around, all sorts of shape, size, colour and texture. Even I grew a beard for a while, but it didn't come to much. Eventually the army organized a water supply with lorries which drove up and down all day, bringing water, I don't know from where. This precarious situation persisted until the first thaw, when the pipeline began to function again and most beards vanished.

There were quite a few theoretical physicists on the mesa, and some very distinguished ones. One of the most interesting was a very young man, barely a student, whom everybody recognized as a budding genius, Richard Feynman. He was exceedingly quick and bright and full of mischievous ideas. He worked out how a combination lock functioned, and after that he could open other peoples' safes by just listening to the tiny clicks that could be heard when the knob was turned forward and backward in the prescribed manner; he embarrassed some people either by leaving their safes open so that they were scolded by the security officer, or by leaving mysterious messages inside.

All the children knew a hole in the fence that surrounded the whole establishment. That hole was not very far from one of the official entrances which were of course guarded, and Feynman once amused himself by driving the guard to distraction. He would crawl out through the hole and then walk in at the official entrance, showing his pass. Within minutes he would be there again coming from the outside, having gone out through the hole; the guard couldn't understand how a man could keep coming in and never going out, as the nearest official exit was a mile or more away. Today Feynman is a household name among theoretical physicists, and his brilliant lectures have been used by a generation of physics students on both sides of the Atlantic.

Another impressive man was Hans Bethe. In contrast to the quicksilvery Feynman he looked and spoke somewhat pedantically, like a German professor; in fact he was the son of one. But behind this slow speech, his endearing smile and his ringing laughter, there was a mind of formidable speed and power. During a party I once posed a mathematical puzzle to both of them and was ready to bet my shirt on Feynman, but Bethe was the first with the answer.

Then there was Stan Ulam, a Polish-born mathematician with

35. Stanislaw M. Ulam, a
Pole who settled in the
U.S.A. and, for a time,
applied his great skill in pure
mathematics to problems of
atomic weaponry.

36. Victor F. Weisskopf,
Austrian-born theoretical
physicist, now retired from
the M.I.T. (Massachusetts
Institute of Technology)
after a distinguished career
which included some years as
director of CERN (Centre
Europeen pour la Recherche
Nucleaire) at Geneva.

an attractive French wife. He explained to me that he was a pure
mathematician who used to work entirely with abstract symbols,
but had now sunk so low that his latest report had contained actual
numbers, indeed numbers with decimal points; that (he pretended)
was the ultimate disgrace! In fact he had an uncanny skill of using
the most abstruse and abstract techniques of mathematics for
predicting the behaviour of an atomic bomb.

Vicky Weisskopf I had known for many years. His older brother
had been to school with me and had once invited me to tea because,

he said, his kid brother was asking so many questions about physics which he couldn't answer. Being two years older than Vicky I was then able to answer some of his questions; it was the last time I could feel I knew more physics than he did. He was widely admired, and still is, not only for his deep insight in questions of physics but also for his political skill. For a time he was something like Mayor of Los Alamos, though of course any civic organization had only limited power against the wishes of the army which ran our community. Let me also give you a sample of his wit. Years later we met at a conference and Vicki said to me 'those young men talk a lot of nonsense; it is up to us older ones, with knowledge and courage, to tell them where they get off.' 'But I won't be sure if what they are saying is nonsense' I replied. 'Ah' said Vicki 'that is where the courage comes in!'

On the whole, despite the numerous security rules, we had a fair degree of freedom. There was a great deal of countryside to roam in; many a time I was taken along by friends who drove to the foot of one of the major peaks and with whom I climbed mountains of 14000 feet or more, higher than I had ever been. And there was Santa Fe, the official capital of New Mexico, less than two hours' drive away (even over those rutted roads, at the war-time speed limit of 35 m.p.h.); there was a constant traffic of people who went to town to do some shopping or sightseeing or to dine at the hotel 'La Fonda'. The townspeople wondered what was happening 'on the hill', and wild rumours were rife, probably put about by our own security people: it was a maternity home for female military personnel, a submarine base (in the middle of the continent!!), and so on. One strapping fellow of Los Alamos was buttonholed by an old lady with the words 'and what are *you* doing out of the army, young man?' His staccato reply left her speechless: 'Madam, I am a Japanese spy.'

We had good salaries and no taxes (at least we British didn't pay taxes) and so we had lots of money to spend. Some of it went on booze, and I remember that one morning I caught myself with an empty glass in my hand and realized that I had just poured myself a tequila and drunk it without even noticing what I was doing. It gave me quite a shock and I firmly decided never to drink by myself; I might well have become a drunkard. Much later I learned that it was against the rules to bring liquor to Los Alamos; but everybody did.

Or else one could collect Indian handicraft products, in particular rugs. I found them fascinating, the many different styles and vigorous patterns, in particular the ones that were made entirely from the natural wool of differently coloured sheep. By the time I left I had a pile about a foot high. Pottery I didn't buy much, partly because I realized that when I left it would be tricky to pack. In fact only one piece of the few that I acquired has survived to this day. For the ethnologically minded there was unending fascination in the archaeological sites and the pueblos where the Indians now lived, in some respects much as their forebears may have lived before the Spaniards came, though they had acquired a good many of the implements of modern American life. They still had adobe houses built from dry mud brick, with the roof beams sticking out and giving the houses their characteristic appearance, often with bunches of red peppers hung up to dry. But at the same time the men wore jeans and print shirts bought from mail-order houses, sometimes even during their ceremonial dances, and some of them had motor cars. A good many of the Indians worked as cleaners, janitors and waitresses on the site.

Most scientists worked in groups of about a dozen people, headed by a group leader, but there were some exceptions. For instance, I was not put into any group, but worked for a while as a sort of visiting plumber, going round to see what various groups were doing, and offering advice when required, particularly on questions of instrumentation. On the other hand there was one group with the unusual feature of having two group leaders, the Italian Bruno Rossi and the Swiss Hans Staub. They both respected and liked each other so much that neither wanted to be the boss of the other; so it was agreed that they should be joint group leaders. But they were very different in their temper. Hans Staub, whenever the least bit went wrong, would burst out with 'Cheesus Christ', in his strong Swiss accent, whereas Rossi only in the most catastrophic circumstances might give vent to the exclamation 'Oh dear'.

Another group leader who comes to mind is John Manley. He had created much of the early organization of Los Alamos but he appeared to have assembled his own group on the principle that every member had to be keen on both music and Chinese cookery. One of them was Ted Jorgensen, who owned the one good grand piano I knew on the site, and moreover kept it well tuned. He also had a considerable knowledge of Chinese cookery which he had acquired five years previously by having a Chinese student for a lodger. When that student left he bequeathed the Jorgensens a cookery book printed in Chinese. Undaunted, they got a Chinese dictionary and proceeded to cook their way through the book. In Los Alamos they gave a dinner party every few weeks on Sunday evenings. I still remember the pleasure of coming home from skiing and, after a shower and a change, to wander over to the Jorgensens where I started to play on their excellent piano while the younger members of Manley's group were cutting up the meat and vegetables in the kitchen. After a while a violinist would arrive and join me in a sonata which we broke off to switch to a trio as soon as the cellist appeared. At exactly eight o'clock the kitchen door opened, and a procession emerged carrying the first two platters; we stopped our music and fell to. After half an hour the crew went back to the kitchen, the musicians to their trio. That was repeated three or four times, and the party ended around eleven o'clock, with well-filled stomachs and ears full of Beethoven and Mozart.

Another culinary occasion was the roasting of shish kebab. I got into this because on one occasion I was asked to let one of my assistants, who was skilled at iron-work, make a set of spits for roasting those little lumps of meat, pickled in wine and interlaced

with tomato and onion, over a charcoal fire. In return I was invited
to the party and waxed so eloquent in my praise of that delicacy
that from then on it was unthinkable that anybody should serve
shish kebab without inviting me. Altogether I had quite a lively
social life and, being an absent-minded person, I was once reduced
to the expedient of sending a message over the paging system 'Will
the person who invited Otto Frisch to dinner please phone the
number of his room'. I don't remember whether that emergency
call had the desired effect.

My work consisted mainly of small projects, such as developing
bits of instrumentation. But there was one piece of work which was
of more than normal interest. When I made the proposal for that
experiment it was sent – as usual – to a group of high-ranking
physicists who decided how the means of the site had best be
employed; I didn't really expect the proposal to be accepted and
was surprised when it was. It was nicknamed the 'Dragon
Experiment', because Richard Feynman (who despite his youth
was on the council) had received it with a chuckle and the words
'like tickling the tail of a sleeping dragon'. The idea was that the
compound of uranium-235, which by then had arrived on the site,
enough to make an explosive device, should indeed be assembled
to make one, but leaving a big hole so that the central portion was
missing; that would allow enough neutrons to escape so that no
chain reaction could develop. But the missing portion was to be
made, ready to be dropped through that hole so that for a split
second there was the condition for an atomic explosion, although
only barely so.

Of course they quizzed me what would happen if the plug got
stuck in the hole, but I managed to convince everybody that the
elaborate precautions, including smooth guides and careful checks
on the speed of each drop, would ensure complete safety. As the
result I became a group leader for a while, and with the help of
a dozen keen collaborators the experiment was set up and completed
in a matter of weeks. It was as near as we could possibly go towards

starting an atomic explosion without actually being blown up, and the results were most satisfactory. Everything happened exactly as it should. When the core was dropped through the hole we got a large burst of neutrons and a temperature rise of several degrees in that very short split second during which the chain reaction proceeded as a sort of stifled explosion. We worked under great pressure because the material had to be returned by a certain date to be made into metal and assembled to form a real atom bomb. During those hectic weeks I worked about seventeen hours a day and slept from dawn till mid-morning.

Other experiments with this atom bomb material were less flamboyant but in fact more dangerous. One man was killed by a runaway reaction while I was there, and later my successor, Louis Slotin – very likeable and popular – was the victim of the second accident. The danger was psychological; assembling a mass of uranium-235 was something we completely understood, and as long as we hadn't reached the critical amount – when the chain reaction began to grow spontaneously – the assembly was completely harmless. But critical conditions could be reached very suddenly as the result of a minor mistake.

Most of the assemblies we made were meant to help us find out exactly how much material would be needed for the bomb; they had neutron-reflecting material round them and were assembled from small bricks, with the reflector in the form of somewhat bigger bricks made from non-fissile material. We had made a strict rule that nobody was to work all by himself, and that nobody should ever hold a piece of material in such a way that if it dropped it might cause the assembly to become critical. The first man to be killed, Harry Daghlian, had broken both rules. He was so eager that he wanted to do one more assembly after everybody else had gone, and a large chunk of heavy metal slipped out of his fingers, and fell on the almost completed assembly; even as he instantly swept it aside with a blow of his muscular arm he saw a brief blue aura of ionized air around the assembly. He felt nothing else but became sick as the ambulance (he had reported by phone) took him to the hospital; two weeks later, his blood count way down, he died from some trivial infection which his body could no longer fight. Louis Slotin, I was told, lived only nine days after a pencil he had placed under a piece of reflector slipped out.

How did an experienced and cautious physicist like Louis Slotin make such a stupid and fatal mistake? Did he really think a pencil was a safe support for a critical piece of material? Or did something deep down in his mind tempt him to play atomic roulette? We shall never know.

There was one occasion when I myself very nearly became victim of a similar accident. We were building an unusual assembly, with no reflecting material around it; just the reacting compound of uranium-235, because this was a good way to test the reliability of our calculations. For obvious reasons we called it the Lady Godiva assembly. I did follow all the rules. I had a student helping me. He was standing by the neutron-counting equipment, and we both watched the little red signal lamps blinking faster and faster and the meter clattering with increasing speed. Suddenly to my surprise the meter stopped; I looked up and saw that the student had unplugged it. Immediately I leant forward and called out 'Do put the meter back, I am just about to go critical', and at that moment, out of the corner of my eye I saw that the little red lamps had stopped flickering. They appeared to be glowing continuously. The flicker had speeded up so much that it could no longer be perceived.

Hastily I took off some of the blocks of the uranium compound which I had just added, and the lamps slowed down again to a visible flicker. It was clear to me what had happened: by leaning forward I had reflected some neutrons back into Lady Godiva and thus caused her to become critical. I hadn't felt anything, but after we had completed the experiment with appropriate care, I took a few of the blocks and checked their radioactivity with a counter. Sure enough, the activity was many times larger than what should

37. William Penney, British mathematician whose lecture at Los Alamos on bomb damage inspired this sketch by the author. He later became chairman of the U.K. atomic energy authority and was raised to the peerage in 1967.

have accumulated if that little incident hadn't occurred. From that I could calculate that during the two seconds while I was leaning over the assembly the reaction had been increasing, not explosively but at a very fast rate, by something like a factor of a hundred every second. Actually the dose of radiation I had received was quite harmless; but if I had hesitated for another two seconds before removing the material (or if I hadn't noticed that the signal lamps were no longer flickering!) the dose would have been fatal.

After the uranium-235 was sent back to be turned into atomic bombs I was a bit in the doldrums. Some of us were sent to distant places to supervise the construction of a variety of components and ultimately the assembly of the first bomb to be exploded in a piece of desert over a hundred miles from Los Alamos. But I was not a good organizer, and everybody knew it; so I was not called upon, neither then nor when finally the real weapons were assembled and taken to be dropped on Japan. Instead I was one of the many people who prepared various experiments to be carried out near the first test explosion. The particular experiment which I had attempted was not very important; moreover it failed because I had underestimated the amount of radiation which would be around, and my film (meant to take a picture of the explosion by its own X-rays) was completely blackened.

Early in July we all went in cars and buses to the test site, code-named 'Trinity', in the desert near Alamogordo, also known as El Jornado del Muerte, Spanish for the Journey of Death. It was not real desert, but very dry country with cacti and other sparse vegetation and a startling variety of arthropods, some of them said to be unpleasantly poisonous. I saw a tarantula which someone had captured and kept in a large jam jar, just about large enough to hold that huge spider. We lived in big tents and for varying times (about a week in my case) while all the preparations were made. A steel tower, about a hundred feet tall, had been constructed to carry the explosive device (not really a bomb, without a streamlined case). When it finally arrived and was being hoisted to the top I was standing there with George Kistiakowski (our top expert on explosives), at the bottom of the tower. 'How far away', I asked him 'would we have to be for safety in case it went off?' 'Oh', he said, 'Probably about ten miles.' 'So in that case', I said, 'we might as well stay and watch the fun.' Actually there was no real danger as the trigger mechanism was not yet armed; that was left to the last moment.

When finally the day came the weather had changed; there were thunderstorms in the vicinity. We had reason to fear that lightning might set off the explosion prematurely, and many of the measurements would be ruined if the weather wasn't clear and quiet. So we had to wait. Some of the big shots went to a bunker only about ten miles away, but most of us were taken to a point in the open, twenty-five miles from the explosion site.

There we sat around through the night, waiting for the weather to clear up. For some hours I dozed in the car, waking up whenever the loudspeaker said something (in between it was playing dance music). Finally the announcement came that the count-down was beginning; now it would be only minutes before the explosion took place. By that time the very first trace of dawn was in the sky. I got out of the car and listened to the count-down, and when the last minute arrived I looked for my dark goggles but couldn't find them. So I sat on the ground in case the explosion blew me over, plugged my ears with my fingers, and looked in the direction away from the explosion as I listened to the end of the count. . . five, four, three, two, one. . .

And then, without a sound, the sun was shining; or so it looked. The sand hills at the edge of the desert were shimmering in a very bright light, almost colourless and shapeless. This light did not seem to change for a couple of seconds and then began to dim. I turned round, but that object on the horizon which looked like a small sun was still too bright to look at. I kept blinking and trying to take looks, and after another ten seconds or so it had grown and dimmed into something more like a huge oil fire, with a structure that made it look a bit like a strawberry. It was slowly rising into the sky from the ground, with which it remained connected by a lengthening grey stem of swirling dust; incongruously, I thought of a red-hot elephant standing balanced on its trunk. Then, as the cloud of hot gas cooled and became less red, one could see a blue glow surrounding it, a glow of ionized air; a huge replica of what Harry Daghlian had seen when his assembly went critical and signalled his death sentence. The object, now clearly what has become so well known as the mushroom cloud, ceased to rise but a second mushroom started to grow out from its top; the inner layers of the gas were kept hot by their radioactivity and, being hotter than the rest, broke through the top and rose to even greater height. It was an awesome spectacle; anybody who has ever seen an atomic explosion will never forget it. And all in complete silence; the bang came minutes later, quite loud though I had plugged my ears, and followed by a long rumble like heavy traffic very far away. I can still hear it.

Los Alamos 1943–1945: 2

We were very glad when Enrico Fermi came to join us, rather late in 1944. His reputation was formidable. In the 1920s he had created some of the fundamental concepts in theoretical physics; in the 1930s he had initiated experimental research on neutrons in Rome with the sensational results that I mentioned earlier, and after collecting his Nobel Prize in the autumn of 1938 he had not returned to Italy but taken his family to the U.S.A. and turned himself into an engineer. Almost on arrival he heard from Niels Bohr about the discovery of fission and started work, at the Columbia University in New York, aimed at measuring the average number of neutrons set free in each fission process. That number would decide whether uranium was a potential source of energy on a large scale, by making a nuclear chain reaction possible.

A human population is an example of a chain reaction. It increases if more than two children in each family, on an average, reach maturity and again start a family; the birth-rate has to be rather more than two per family since some children die and some never marry. In the same way, in uranium each fission (except the very rare spontaneous ones) consumes a neutron, and unless at least one fresh neutron is thrown out from the fission fragments there could be no sustained chain reaction. Even when – both in France and in the U.S.A. – the average number was found to be much larger, between two and three, it was not yet certain that the 'mortality' of neutrons, the risk of being caught without causing fission, could be kept small enough to allow on the average at least one neutron per fission to survive and cause another fission.

One of the main risks to a neutron comes from uranium itself, from the main isotope uranium-238 which suffers fission only from the impact of an exceptionally fast neutron; slower ones just get swallowed. But the comparatively rare isotope uranium-235 can be split even by slow neutrons and moreover is very good at catching them; indeed various physicists realized early on that the best chance of getting a chain reaction was by combining uranium with a 'moderator', a material that slows neutrons down. Water is a good moderator, as Fermi had found in 1935; the protons (nuclei of hydrogen-1) which it contains rob a neutron of its energy in a dozen collisions. But they also tend to capture a neutron, forming nuclei of hydrogen-2, also called heavy hydrogen. Hence water is not a good enough moderator to give a chain reaction when combined with natural uranium.

But water does contain some nuclei of heavy hydrogen, which are not interested in catching yet another neutron, and by rather expensive processing one can concentrate them and make heavy water, which is an excellent moderator. Chain reactors employing heavy water have since been built. But Fermi decided to use carbon, in the form of graphite; a neutron needs more collisions with those heavier nuclei to lose its energy, but fortunately the chance of being captured in such a collision is very small. Under Fermi's guidance (he was an 'enemy alien' but the famous American physicist Arthur H. Compton who was in charge gave him every support) the work was pursued with the utmost urgency, particularly after Pearl Harbor which brought the U.S.A. into the war. Many tons of specially pure graphite were obtained and tested, and almost exactly one year later, on 2 December 1942, in a squash court under the stands of a Chicago football ground known as Stagg Field, the first nuclear chain reaction was achieved. Its control, by moving neutron absorbers made from cadmium, turned out to be quite easy.

Why that tremendous urgency? Not just to prove that a nuclear chain reaction was possible, opening the door to the storehouse of nuclear power. No, the main reason was that the chemists in California had obtained barely visible amounts of a transuranic element, later to be called plutonium, formed when neutrons were captured by nuclei of uranium-238. It was expected, and soon proved, that this would react to slow neutrons rather like uranium-235 and hence be a possible material for atomic bombs, if one could make enough of it.

Even in Fermi's first chain reaction only small amounts of plutonium-239 would be formed; but when that pilot model was found to work, the construction of very much larger piles of uranium and graphite was begun, capable of making this alternative bomb material in kilogramme amounts, in case the separation of uranium isotopes proved too difficult. In the end both methods were made to work at the same time; of the first two A-bombs used in anger – over Hiroshima and Nagasaki – one contained uranium-235, the other plutonium-239.

Those production piles were built by the Du Pont organization, near a hamlet called Hanford, not far from Seattle, and Fermi assisted with the design. For security reasons he was known as 'Mr Farmer' and was given a bodyguard. On the long walks which were his recreation he liked to talk physics to his guard. One day he said pensively – he had acquired a good command of English but had never lost his Italian accent – 'my bodyguarde, now he know so much physics, he will soon neede bodyguarde himself'.

Fermi never appeared to be in a hurry, yet he got a great deal done because he was well organized. He soon realized that people tended to come and interrupt his work by asking him questions, so he announced that he would be in his office all morning, available to anybody who cared to come. In the afternoon he would wear his laboratory coat and work in the laboratory, unavailable. On Sundays he was usually out walking with a group of young people who felt entirely at ease with him though he was obviously the master. I have never met anyone who in such a relaxed and unpretentious way could be so completely dominant.

Niels Bohr also came to Los Alamos. He arrived with his son, Aage, and they were to be known as Nicholas Baker and Jim Baker respectively. Their arrival was surrounded with such fuss and mystery that one of my friends asked why they hadn't been packed and sent in a crate; it would have been so much simpler. In the autumn of 1943 Niels Bohr had escaped from Denmark when he and his family crossed the sound to Sweden in small boats on a dark night; word had gone round that the Germans were about to pick up and imprison all the Jews in Denmark, and Bohr had a Jewish mother. On arrival he immediately travelled to Stockholm and managed to get an audience with the King of Sweden whom he persuaded to offer asylum to all Danish Jews, by publicly declaring that any that felt in danger would be welcome in Sweden. That broadcast was a gesture of some significance; it made it easier for the refugees, who were picked up and brought to safety on reaching Swedish territorial water, and perhaps persuaded other governments to be more generous in granting refuge.

Bohr was then invited to England by Winston Churchill. He was taken in a Mosquito bomber, seated in the bomb bay, where he became unconscious because the headphones didn't fit his very large head and so he didn't hear when the pilot told him to turn on the oxygen. His silence worried the pilot, who could not turn back. But once he had crossed Norway he descended to lower altitude and Bohr revived.

Niels Bohr got in touch with statesmen, such as Churchill and Roosevelt, whom he tried to convince that the principles of the atomic bomb should be made public rather than kept as a secret which would anyhow very soon be broken; a prediction in which he was completely right. He tried to warn the politicians against the dangers of an atomic arms race; but his foreign enunciation and his habit of slow and patient argument probably irritated the men whose habit of quick and bold decision had placed them where they were and to whom the immediate goal of winning the war was paramount. Later, after the war, he addressed an Open Letter to the United Nations, again calling for an open world without atomic 'secrets'. I wonder how many people have studied that long document (around 6000 words) with the care it deserves.

But during his visit in Los Alamos he was in a relaxed mood. He could see that the work was going well and the war was turning in our favour. The radio reports were confusing, but Bohr always listened; 'we must hear all the rumours before they are denied' he joked. Until the bomb had shown its power no political action would be possible. The weather was fine as usual, and on one occasion it was my turn to go for a walk with Bohr. But I had forgotten one of the elementary rules: I had no matches. So when Bohr pulled out his pipe and asked me for a light I had to disappoint him. But I offered to light his pipe for him, having noticed that he was long-sighted and wore quite strong lenses. In the intense sunshine of New Mexico it was not difficult to use one of his lenses as a burning glass and set fire to the tobacco in his pipe; but while Bohr was intrigued I don't think he was satisfied. To him the lighting and blowing out and throwing away of matches was part of the enjoyment of his pipe, which kept going out because he kept talking.

I once asked him how much he was troubled by his false name; did he always remember to sign himself Nicholas Baker rather than Niels Bohr? His reply was 'What is the difference? My signature is quite illegible; it could be either.' We all enjoyed the story of his arrival in the U.S.A.: he was hustled from the boat to his hotel by a couple of F.B.I. men who were carrying his and his son's luggage, first into a taxi, and then through the reception into the hotel room where they all collapsed with deep breaths of relief that his identity had been successfully concealed. And then one of them noticed that the name NIELS BOHR was written in large black letters on his suitcase.

The Bohrs were not the only ones with false names. Fermi was officially referred to as 'Mr Farmer', and even I changed my name, in a sense. Within a couple of days of my arrival in Los Alamos someone asked me 'What's your first name?' By that time I had discovered that the place was lousy with people called Robert, such as Oppenheimer, Bacher, Serber and a number of others, so on the spur of the moment I spoke the truth and said 'Otto', although I had never been called that in my life. My mother for a while used both my names, and some of my relations even spelt them as one, Ottorobert, not uncommon with German names. But there it was: now I was known as Otto to all my American friends, and still am; though over the public address system the name sounded ominously like Adolf. When Los Alamos got a small radio transmitter going I was given a weekly slot playing the piano for fifteen minutes as 'our pianist'; even the name Otto was thought to give away too much to a passing motorist.

Life in Los Alamos was, as I said, not all physics. When plans were announced to produce a play, the well-known *Arsenic and Old Lace*, everybody was invited to audition for it, but I didn't bother; with my strong foreign accent I wouldn't have a chance. Later I was annoyed to find that the play actually called for a foreigner in its cast, and that the role was given to a hundred-per-cent American who tried very hard, and not very convincingly, to sound like a foreigner, which would have been no trouble to me!

But I got my chance at the footlights after all. The producer had the amusing idea that at the end, when all the actors took their curtain calls, the corpses should appear as well; in the play, as you probably know, it is revealed that a number of people had got killed before the curtain rises. So a dozen or so of the best-known people in Los Alamos were invited to sneak out at the end of the play, get a hasty dabbing of flour to make them look like corpses and then join the actors on the stage. There was Oppenheimer and (I think) Teller and lots of others, and I too was one of the corpses. But while everybody else just went on to the stage and bowed somewhat sheepishly, some with grins on their faces, I decided to play it realistically and staggered on stage in a contorted posture, with my hands hanging limply from the crooked arms, my head turned upwards and my teeth bared. All my friends assured me that I had been the best corpse of all.

Another social occasion I remember was a circus, organized for the benefit of the children. There were lots of acrobatics and so on, mostly by the children themselves, but some grown-ups were also roped in. One lady undertook to read the future in a crystal ball; she wore an impressive gipsy costume and played her role very convincingly. I volunteered as a lightning artist and was looking forward with fiendish delight to drawing particularly malicious caricatures of all my friends, and taking their money to boot! For the occasion I had dressed up in a velvet jacket and a flowing tie and had taken my seat at a little table with a pad of paper and pencil; but to my surprise my clients were all children! I had never drawn a child before and hastily had to work out some trick of drawing a pretty girl or a handsome boy; likeness didn't matter very much. The children were all pleased with their portraits and kept coming in an unending stream; some of them even came twice and paid their quarter dollar a second time. Soon I ran out of paper and pencils and had to organize a shuttle service to take away the small cash, to sharpen my pencils and to bring fresh paper. I never dreamed I would be such a success (I earned about a hundred dollars towards the expenses of the circus) but was robbed of the pleasure of disfiguring my friends.

It was in Los Alamos that I learned to drive a car. I had earned enough money to buy one, a big secondhand Hudson, and started getting lessons from Harry Daghlian. Later, when he had fallen victim to that atomic accident which I mentioned, Louis Slotin took over; it is an odd coincidence that he too was to be killed by radiation from a runaway assembly. When I thought I had had enough lessons I presented myself for a driving test to one of the army officers who thereupon issued me a most peculiar driving licence: it did not contain my name. That was of course done for security reasons; if I was involved in an accident away from Los Alamos my name should not become known.

The day after my test I had my first accident. I was driving three other people down into a canyon and taking an S-bend on loose gravel rather too fast. I suddenly saw a tree in front of me, and, unable to move, saw it come nearer and nearer until it hit my bonnet. Actually the whole thing had probably taken only half a second, too little to take any effective action. The two people behind me were unhurt and helped me out of the car since both front doors were jammed, and Perc King who was sitting next to me had been knocked out when his head hit the windscreen. (He is still alive and we are still good friends.) I had been doing only about twenty miles per hour, but hitting a tree even at such low speed means a rather violent impact, and there were no seat belts in those days. Apart from various cuts and bruises (my knee went through the dashboard and my head broke the rear-view mirror, while my chest bent the steering wheel) I had no serious injury, but spent a few days in hospital, my broken rib fixed with tape. Then I got bored and just went home to the Big House. The main nuisance was that for several weeks whenever a party was getting hilarious I had to leave in a hurry because laughing was very painful.

The Hungarian-born mathematician John von Neumann came to Los Alamos for visits, and since he also stayed at the Big House, we sometimes walked home together. On one such occasion, knowing that he was concerned with the problem of doing complex calculations – the word computing had not yet gained general currency – I asked him whether anybody had thought of building a machine which calculated with the help of radio tubes instead of cog wheels. He seemed interested and invited me to take part in a small discussion to which he had also invited a couple of electronic engineers. Together with a large dog we went into a small office where he began to explain his ideas. Starting about five o'clock, he spoke without notes and, apart from answering occasional questions, didn't stop until ten o'clock, when somebody

38. John von Neumann, Hungarian-born mathematician, who invented the theory of games (including economics and politics) and was the driving mind behind the first electronic computers.

pointed out that if we wanted a glass of beer we had to go before the pub closed. After that glass of beer we went back to the lecture room and continued until midnight. At the end of this my head was buzzing, but I felt I really understood the new world of electronic computers.

It was a fantastic performance on the part of John, who was of course the spiritual father of electronic computers, had one under construction in Princeton at that very time, and had no doubt invited me largely out of kindness and perhaps in the faint hope that I might contribute a new idea, which I didn't. But I never ceased to be fascinated by electronic computers, and I feel that I have been privileged in having been initiated so marvellously by the Master himself. His mathematical achievements are far too subtle and technical for me to understand or to describe, but I can attest to the strength of his brain because I once saw him, for a bet, drink sixteen martinis in a row and then be still on his feet and quite lucid, though somewhat pessimistic in his utterances.

Von Neumann was one of that galaxy of brilliant Hungarian

39. Edward Teller, the Hungarian-born 'father of the H-bomb', but also a theoretical physicist of great versatility and imagination.

expatriates who caused my eccentric friend Fritz Houtermans to propound the theory that these people were really visitors from Mars; for them, he said, it was difficult to speak without an accent that would give them away and therefore they chose to pretend to be Hungarians whose inability to speak any language without accent is well known; except Hungarian, and those brilliant men all lived elsewhere.

Another remarkable Hungarian was Edward Teller whom I had briefly met in Denmark. He had the sort of brain that is always looking for a challenge: a difficult problem, an argument, a game of blindfold chess. Once when we travelled together and I was too tired to argue I asked him if he knew anything about four-dimensional regular polyhedra. He found that problem intriguing, so I was able to doze in peace. Two hours later, as we got off the train, he had achieved about as much understanding of those polyhedra as I had done in three weeks at the age of fifteen when I was probably at my brightest. I liked and admired Teller not least for his piano-playing, where he replaced by musicality and sheer will power what he lacked in technique.

Teller had lost a leg in a traffic accident while young, but he never let on. His artificial leg merely gave him a slight limp, and he didn't mind going for long walks; he even often joined us when we went skiing, but then wore snow shoes rather than skis. He was a fanatical anti-Communist, having seen his native Hungary taken over, and he felt strongly that the U.S.A. must be armed for what he saw as the inevitable conflict. The A-bomb, deriving power from the fission of uranium or plutonium nuclei, was for him only the beginning; the fusion of deuterons (the nuclei of heavy hydrogen, hence the name H-bomb) would allow much bigger explosions, and Teller put his powerful will behind that development. Weight for weight, uranium and deuterium contained about the same energy, but in uranium (as in plutonium; the two behave similarly) only a modest fraction of the energy could be set free, and only a limited mass could be used, or it would have instantly blown up. With deuterium, both those limitations were absent, but it had to be heated to an ignition temperature of millions of degrees by an A-bomb, acting as a mere detonator. The difficult technical problems needed several years, and I did not take part in that work and know little about it; the crucial details are still secret. H-bombs have now been exploded with a thousand times the power of the A-bomb that flattened Hiroshima.

After the experimental explosion at Alamogordo the work of Los Alamos was largely done. Those who were good organizers were sent away to distant parts to arrange the assembly of all the components and the air-lift of the atomic bombs. I stayed behind with most of the others, and there was comparatively little to do, apart from documenting the work we had done. But how to use the new weapon soon became the subject of lively discussions. Should it be used at all? Should a demonstration be staged on an uninhabited island with the enemy invited to watch it? Much debate went on at various levels; among politicians, among the big shots at Chicago such as James Franck and Leo Szilard, and even to a smaller extent at Los Alamos itself. I was never very politically minded, as I have said before, and I had an extra reason not to take part; word had come down from the British Government that we as guests in the United States should keep out of political discussions. So while I heard a certain amount, I had a good excuse for keeping my mouth shut.

Some of us said that scientists ought to put their weight behind what they felt to be the right course of action; others took the line that the cobbler should stick to his last. I remember a story being told about the Greek sculptor Phidias who had completed a new statue of Zeus and hidden behind it to hear what the passing Athenians said. When he heard a cobbler say 'The big toe is too large' he came back later at night and chipped a bit off the big toe. The next morning he saw the cobbler pass again, remarking that the toe had been improved but the elbow wasn't right. At that, Phidias stepped out from his hiding place and addressed the cobbler with the words, 'When you talk about toes you talk about what you know, and I listen; but I pay no attention when you talk about elbows.' The moral being, of course, that scientists should stick to matters of their own competence, and at the time I found that view very plausible. I am no longer convinced that this is always right. Scientists are trained to think objectively and dispassionately, an asset for making decisions of any kind.

We didn't know when the bomb would be dropped in earnest or where it would be dropped. Then one day, some three weeks after Alamogordo, there was a sudden noise in the laboratory, of running footsteps and yelling voices. Somebody opened my door and shouted 'Hiroshima has been destroyed!'; about a hundred thousand people were thought to have been killed. I still remember the feeling of unease, indeed nausea, when I saw how many of my friends were rushing to the telephone to book tables at the La Fonda hotel in Santa Fe, in order to celebrate. Of course they were exalted by the success of their work, but it seemed rather ghoulish to celebrate the sudden death of a hundred thousand people, even if they were 'enemies'. On the other hand there was the argument that this slaughter had saved the lives of many more Americans *and* Japanese who would have died in the slow process of conquest by which the war might have had to be ended had there been no atom bomb. But few of us could see any moral reason for dropping a second bomb (on Nagasaki) only a few days later, even though that brought the war to an immediate halt. Most of us thought that the Japanese would have surrendered within a few days anyhow. But this is a subject that has been endlessly debated and never settled.

Research resumed

Had the war been a disaster for physics by dragging so many of the best men away from their benches? Indeed not; we had forged new weapons, not only for war but also for our own armoury. Once they were deployed, progress became spectacular; I should guess that the time lost by the war was made good within a few years.

The first large nuclear reactors had been built to manufacture the nuclear explosive, plutonium. But they also produced vast numbers of neutrons, potentially dangerous and confined inside thick concrete walls; by leaving a small opening one could get beams of neutrons and study many of their properties with ease and precision.

In particular their selective absorption in nuclei which we had studied laboriously by estimating their energy from their absorption in boron could now be observed by selecting neutrons of definite energy, by much the same method that Stern and I had used in Hamburg. A fast-spinning wheel with slots allowed neutrons to escape from a reactor in brief bursts, and their arrival at a counter several feet away was accurately timed. The densely spaced energy levels which Niels Bohr had predicted could now be seen in detail, and even for particular isotopes, separated by the methods which had been developed for separating the isotopes of uranium.

Short time intervals between electric pulses could be measured by methods developed for radar, which obtains the distance of an aeroplane by timing a brief pulse of electric waves reflected from it. Those methods have since been greatly refined, and a microsecond (one millionth of a second) is now too large a unit; one thousandth of it, one nanosecond – a time in which light travels about one foot – is more appropriate. By timing a particle – by measuring its time of flight – over a distance of a hundred feet we can obtain its speed to 1% even if it flies almost as fast as light; and no particle has ever been seen to defy Einstein and exceed that ultimate speed limit. The speed of a particle together with the way its path is bent in a magnetic field tells us its mass. This 'time-of-flight' method has been a great help in identifying the many new particles which are created in violent collisions; more about that later.

Another spin-off from radar was the birth and growth of an entire new science: radioastronomy. But I shall leave that to the last chapter as I was able to watch part of that fascinating story right here in Cambridge.

In nuclear physics, soon after the war, there was an event seemingly more suited for the social column: the unexpected return of the scintillation screen, that Cinderella of nuclear physics ever since electric counters ousted her in the thirties, through her sensational marriage to the photomultiplier, prince of the television age. In more sober terms, every electron released in a photomultiplier by a quantum of light is multiplied a millionfold by hitting several electrodes in succession; each impact releases several new electrons, forming a huge avalanche in a few nanoseconds. A photomultiplier can 'see' the flashes caused by fast electrons in a scintillator, and can even measure the brightness of those flashes, too weak and diffuse to be seen by the human eye.

Made by several people independently, this was a most timely invention; Geiger counters, limited to a few thousand particles a minute, could not cope with the vast amounts of radioactive materials now available. The new scintillators could count a thousand times faster; moreover, the brightness of each flash was a fair measure of the energy of each electron. Instead of thin screens, inch-size blocks of transparent materials could be used for efficiently counting gamma rays which mostly go clean through Geiger counters. Naphthalene (the stuff of moth-balls) was one of the first scintillators used, soon to be replaced by specially designed scintillating plastics and liquids.

So much for this new and powerful technique; what about results? Let me start with the most important single experiment: the final proof (1956) of the existence of neutrinos, those very elusive particles which Pauli had dreamt up in 1930. People were then worried about the 'beta process' whereby nuclei can alter their charge by emitting an electron: Peter Debye called it 'a subject best not thought about, like the new taxes'. The problem was that the individual electrons emerged with widely varying energies; yet, for a given radioactive isotope, they emerged from the same kind of nucleus and left behind the same kind of final nucleus. No reason for that random variation had been found in fifteen years of search.

For a while Niels Bohr seriously considered that perhaps the law of the conservation of energy did not hold in beta processes. Bohr felt that every advance in understanding had to be paid for, usually by giving up some previous 'certainty'. To understand planetary motion, Copernicus had to give up the belief (as Aristarch had indeed done 1800 years earlier, but that was forgotten) that the Earth was the unmoving centre of the world; Einstein gave up universal time in order to reconcile mechanics and electrodynamics in his relativity theory. Was it now time to sacrifice energy conservation?

Pauli had a different idea. In a famous letter, starting with 'Dear radioactive ladies and gentlemen...' (Lise Meitner was one of the addressees) he suggested that with the electron another particle was emitted, the two sharing the available energy at random. That second particle had to be without electric charge and very light and Pauli called it a 'neutron'; when Chadwick discovered his much heavier neutron, Pauli's neutron was renamed neutrino (little neutron) by his Italian colleagues, and that name has stuck.

Enrico Fermi developed Pauli's idea into a proper theory, and its conclusions were confirmed by experiments. But all attempts to catch neutrinos ended in failure; clearly they were not affected by electromagnetic forces or by the even stronger forces which held nuclei together. One thing was sure: a neutrino must be affected by the force that was able to cause its creation; but that force was clearly very weak since nuclei took hours, days or years to perform that feat, immensely long times on a nuclear scale. It was calculated that a neutrino could traverse the whole earth and emerge at the antipodes, with hardly a chance of being stopped by a nucleus on the way. Obviously very strong neutrino sources and very bulky detectors were needed to catch that slippery customer.

Nuclear reactors offered the one, and scintillation counters the other. In 1953 a tank with over seventy gallons of scintillator fluid, stared at by ninety photomultipliers, was placed near one of the largest reactors in the U.S.A. at the initiative of Frederick Reines. The expected signals were found, but there was still doubt; in 1956, an improved experiment brought certainty. Of the vast number of neutrinos that escaped from the reactor – about 10^{20} every second – one was caught every twenty minutes. Not much, but enough to carry conviction, and indeed just as many as predicted; and in time to please Pauli, who died a couple of years later.

That was the single most striking achievement of the powerful new technique of counting scintillations with photomultipliers. But that technique also produced abundant new information about nuclear energy levels. Regularities which had been dimly guessed could now be clearly seen; they pointed to a structure within nuclei, comparable to the shell structure of the atomic electrons, elucidated in the twenties. Had Niels Bohr then been wrong in likening a nucleus to a drop of liquid? Surely nucleons could not orbit in such crowded conditions?

Once again Pauli holds the key, with his exclusion principle. I like to imagine a polite exchange between two nucleons: 'We seem to be headed for a collision, but I can't go into a different orbit, nor can you; Pauli says the low ones are all in use, and we lack

40. Aage Bohr, son and successor of Niels Bohr; a leader in the theory of atomic nuclei, he shared the Nobel Prize with Ben Mottelson in 1976.

the energy to go into a higher one. So we must carry on as if we had not collided; see you again!'

Whatever you think of that conversation, the model worked and accounted for a large number of facts. It explained the 'magic numbers' 8, 20, 28, 50, 82 and 126: nuclei containing just that many neutrons or protons had long been known to be particularly stable, and exactly those numbers could be deduced from the shell model. It also accounted for the spins and magnetic moments of nuclei, studied by Kopfermann in Copenhagen and others elsewhere. The nuclear shell model was invented independently by the German-born Maria Goeppert-Mayer in the U.S.A. and by Otto Haxel, Hans Jensen and Hans Suess in Heidelberg; Jensen and Goeppert-Mayer shared half the 1963 Nobel Prize.

Yet the liquid-drop model was not dead; it was still the best for describing a highly excited nucleus, with the electrons behaving more like a milling crowd than like a polite dance. But to understand the intermediate region of nuclei in their lower quantum states required a rare combination of quantum-mechanical intuition and mathematical skill. It was Niels Bohr's son and successor Aage (rhymes with 'saw a'; you don't hear the g) and Ben Mottelson who achieved that in their 'collective model', a quantum-mechanical description of nucleons orbiting inside a wobbly rotating droplet. Their joint work, extending over a quarter of a century and accounting for many fine details of nuclear behaviour, made Margrethe Bohr in 1976 the mother as well as the widow of a Nobel Prize winner; my affectionate thoughts go out to her, still as kind and charming as ever at eighty-five when I last saw her.

But what did most to catch the public imagination and to change the very style of physics was the development of large particle accelerators, those battleships of physics, literally miles in size and made from many thousand tons of steel and copper. Why do governments allot millions of pounds to build and operate them, with crews of hundreds working round the clock?

It all began with the cyclotron, invented and built by Ernest O. Lawrence and M. Stanley Livingston in Berkeley, California, in the early thirties. The basic idea was to accelerate particles not by a single strong shove, using very high voltage, but by many small shoves. A magnet a few feet in size was used to save space by making the protons run in circles. As the proton gained speed the circle expanded in direct proportion so that each circle took the same time; a strong electric field of the right radio-frequency produced regular shoves, two for each turn, and with a hundred shoves the protons could be given energies of about 10 MeV, enough to split even heavy nuclei; deuterons (nuclei of heavy hydrogen, discovered about that time) were even better bullets.

Lawrence was a bold experimenter and often got away with the 'impossible'. Relativity theory imposed limitations; but when Bethe computed them Lawrence had already beaten them by making the magnetic field less regular; why that should help was only understood later. He had simply told his technician to push pieces of iron sheet into the magnetic field; to the report 'No good, boss; the beam got worse' he replied 'Excellent; if it can get worse it can get better!' And it did, with some trial and error. During the war he built large magnets for use as mass spectrometers to

41. Ernest O. Lawrence (1901–1958) the American physicist who first (in 1930) accelerated particles to high energies in many small steps, both in a straight line (linear accelerator) and in a spiral (cyclotron); Nobel Prize 1939.

separate uranium isotopes; when copper supplies for magnet windings ran short tons of silver, an even better electric conductor than copper, were borrowed from the U.S. Treasury.

After the war the relativistic barrier was shattered, by varying first the radio-frequency while the particles gained speed, and later the magnetic field as well. All kinds of ingenious tricks were invented for nudging the particles back to their proper path when they happened to drift away from it. Now the door was open to ever-increasing energies, limited only by the increasing cost of ever larger magnets.

But what was the purpose? Even the heaviest nuclei were defenceless against protons of 20 MeV; surely to go to a hundred MeV or more was just overkill? The answer is that those machines were not meant for atom splitting; they were meant to cast light on the force by which nucleons act on one another. By studying how two protons rebounded from a collision one hoped to find out how that force varied with the distance between the protons. It was really an attempt to extend what Rutherford had done forty years earlier. But it didn't work out that way; there were big surprises ahead.

184

Up to a hundred MeV it all looked gratifyingly simple; the protons behaved more and more like little billiard balls, a ten millionth of a millionth of an inch in diameter. But around 150 MeV the protons appeared to swell; they became more liable to collide with protons in their way, and in some of the collisions new particles, called mesons, were created which had been discovered in the cosmic radiation only a few months previously.

That discovery had not come out of the blue. As early as 1935, the Japanese theoretician Hidekei Yukawa had said that there ought to be quanta related to the nuclear force, just as photons are the quanta related to the electromagnetic force. But whereas photons have no intrinsic mass, his quanta ought to have a mass about 300 times that of an electron. Such a 'meson' ('intermediate' in mass between the proton and the electron) was soon found in the cosmic radiation but did not fit Yukawa's predictions: its mass was only 207 times that of an electron, and its penetrating power was much too large. We now call it a muon and regard it as a heavy kind of electron.

The true Yukawa meson, now called the pi-meson or briefly the pion, was first spotted by the tracks it formed in special fine-grained photographic emulsions. Those tracks, carefully analysed under a microscope, revealed that the pion has an electric charge (otherwise it would form no track at all!) and a mass close to Yukawa's estimate; it is radioactive and transforms itself within a few nanoseconds into a muon. But that is not the end: the muon is again radioactive and explodes, after a mean life of two microseconds, into an electron (or positron, depending on its charge) and two neutrinos.

More new particles were created as larger accelerators were built; several dozens are known today, and new ones are still being discovered. Surely they cannot all be elementary? They remind us of the dozens of different atoms thought to be elementary a hundred years ago, each with its characteristic and unexplained chemical properties. After a hopeful interlude around 1920 when those atoms appeared to have been reduced to just two basic bricks

– the electron and the proton – we seem to be back in the nineteenth century. Several dozen subatomic entities claim the right to be regarded as elementary; are perhaps some of them more elementary than others?

The oldest of all, the electron, still seems to be elementary, and of course its anti-particle, the positron. Being the lightest of all, they are not suspect of hidden complexity, and indeed they obey very accurately the simple equation which Dirac invented in 1928; no test has failed to confirm that. But they are now part of a family, the 'leptons', which includes the muon and two kinds of neutrino, all with their anti-particles. That family of eight keeps itself to itself; only four react to electromagnetic fields, and none to the strong forces which dominate the nuclear world. The proton is certainly not elementary; the first indication came when its magnetic moment, measured by Stern and myself in 1932, turned out to be almost three times as large as expected from Dirac's theory. Its uncharged brother, the neutron, had just been discovered; Pauli had suggested his neutrino, and the positron, predicted by Dirac's theory, was observed in 1933. After that there was a lull, with the basic structure of matter still reasonably simple. The muon, first seen in 1937, was a mere hint of things to come.

The new chapter started when Cecil Powell and Giuseppe Occhialini extended the study of cosmic-ray tracks by using photographic plates, made with very fine grains for that purpose. Coarse tracks formed by alpha particles had been seen early in the century, and attempts to improve emulsions for getting finer detail had been pursued in the twenties by a group of Viennese women scientists (mainly Kara Michailowa, Marietta Blau and Berta Karlik), but then shelved for a couple of decades. Shortly after the war a Canadian chemist, Pierre Demers, made emulsions with much finer grain; the British physicists appointed a committee, headed by my old friend Joseph Rotblat, which got the photographic industry interested in the new possibilities, and a new powerful research tool was developed. Blackett's triggered cloud chamber joined the

42. The author playing the piano at the Michelson Laboratory, China Lake, California, in 1950.

chase, and the scanty supply of particles from the cosmic radiation was enormously increased when the large accelerators came into action.

To trace the detailed sequence would be tedious and confusing; let me briefly state the main facts. The proton turned out to be the lightest member, and the only stable one, of a large and still growing family, the baryons. Some are so short-lived that they can't travel a distance we can measure; we deduce their existence from the particles into which they disintegrate. But the first ones to be seen had lives in the region of nanoseconds, and that was strange because estimates from existing theory gave much shorter lives. So they were called strange particles and were supposed to carry one or more units of something that was hard to get rid of (though not rigidly conserved, like charge or energy); that something was called 'strangeness'.

Research resumed

Have we now gone back as far as the eighteenth century, when scientists said that an object was hot because it contained a fluid called 'caloric'? Not quite, though our ignorance about the nature of 'strangeness' is comparable to their ignorance of the nature of heat. The trouble is that we can't observe the entity which, we think, prevents those strange particles from disintegrating; so we make a joke of it and call it strangeness.

Other properties of particles, such as their charge, mass and spin, can be measured more or less directly, and the results fall into patterns, which we try to analyse mathematically; a bit like a woodworm, watching a game of dice from the ceiling of a pub. He might notice that two white squares with randomly varying numbers of black dots keep appearing on the table; but he would have to be pretty bright to realize that they are the faces of two cubes, never having seen one from close up. We are in a similar situation: we watch varying patterns of mass, charge and spin, and we try to guess at the overall structure of which they might be different aspects.

One interpretation which was out of favour for a while but has come back is the proposal by the U.S. theoretician Murray Gell-Mann that a baryon is made of three 'quarks'. Nobody has yet seen one although – supposedly carrying a charge of one-third or two-thirds that of an electron – they would be easy to detect. They seem to cling very firmly together, or perhaps there is some other reason why they cannot exist separately. (Magnets have two poles, but you never see a single one!) Gell-Mann thought that there should be three different kinds, but that number has gone up lately; varieties are postulated which differ in 'colour' or 'flavour'; even 'charm' is attributed to some, to explain the apparently charmed life of some recently discovered long-lived particles. They are probably not connected with the force of gravity, but in the choice of names I detect the force of levity!

Speculations like that are pursued with great gusto, and it is too early to guess which of them, if any, will survive into the next century. But they are necessary; they predict things that can be observed, and the experimenters are busy testing them, to eliminate some of the speculations and thus allow closer study of the rest. On the whole, experimenters try to measure all they can and often get unexpected results which stimulate new speculations. But probably it will take another Bohr or Einstein to pull all the relevant bits together and create a theory which really makes sense.

Finally the mesons. As I said, Yukawa had predicted them to account for the 'nuclear glue' that holds nucleons together in the nuclei, a strong attractive force at very small distance but totally unobservable at the usual distance between atoms. Given enough energy (about 140 MeV at least) a nucleon can produce a pion (the lightest of the mesons). According to the quantum theory, it can do so on its own if it takes the meson back before the loss can be detected, given Heisenberg's uncertainty principle. The picture that comes to my mind is that of boys wandering about, occasionally tossing a ball in the air and pocketing it again; if two boys get close together they will toss the ball to each other instead and will stay together because they enjoy the game. In the same way nucleons are attracted to one another because when they are near enough they can toss 'virtual' mesons back and forth. It sounds fanciful, but it fits the facts; even the extra magnetism of the proton and the magnetism of the neutron, measured in 1939, fits in with the idea that a nucleon keeps producing mesons and snatching them back.

There are many heavier kinds of meson; they cannot be tossed so far, so they affect the force between nucleons only at still smaller distances. Some are even heavier than protons, defying their appellation as 'mesons'. But we can't keep renaming things; we still speak of atoms ('indivisibles') though they have been cut into many pieces.

Where is this all going to lead? Since I am no Einstein or Bohr

I can't tell you. Probably it will need some new concepts that nobody has thought of yet. And those battleships of physics, the particle accelerators, can be regarded as gigantic microscopes for looking into the innermost secrets of matter. They symbolize the drive and power of pure science, aimed not at material profit but at a deeper understanding of the nature of things.

Return to England

Soon after the end of the war the exodus from Los Alamos began. We had been encouraged to stay and write down the work we had done, in order to preserve and eventually to publish the accumulated knowledge, but that only took weeks. I was looking forward to a holiday in California. Hans Staub, the Swiss who had shared the leadership of his group with Bruno Rossi the Italian, proposed to go back to his house at Palo Alto; he was Professor at Stanford University. He was very attached to California, felt it was the only part of America where one could live, and had, to everybody's amazement, refused an offer of a Professorship in Chicago at something like twice the salary. His opinion was pithily expressed (with a strong Swiss accent) when he stated 'As far as I am concerned, Hitler can have everything east of the Rocky Mountains.'

The transport of his family was no simple matter. His wife with her new baby took a plane. He, with his two school-age daughters, proposed to drive home, and since it was a long drive he invited me to come and take turns at the wheel. I still admire his sang-froid; he actually went to sleep while I – a driver with a pretty poor safety record and no experience at all of city driving – wound my way through Los Angeles. It is a miracle that I got through safely. The last member of the family, the cat, was put into a crate together with some fish and sent by rail. It arrived several days later, starved to a skeleton, the fish stinking to high heaven. Since the cat had been away from their house for about two years, Staub was concerned that it wouldn't know its old home and might run away; we were all told not to let it out of the house.

Well, in the middle of the first night I heard a miaowing, sleepily got out of bed and opened the front door; when a cat miaows you let it out. Then I went back to bed, and in the process I woke up and realized that I shouldn't have. I spent an uncomfortable hour dozing and waking, wondering if the cat was back, if it would ever come back, and what I would say at breakfast if it didn't. From time to time I went to the door and looked outside; no cat. Then, about an hour later, there was the cat on the mat; he came in without a murmur, and all was well. I don't think I confessed.

I stayed with the Staubs for about two weeks. It would have been nice to see a bit of the town and surrounding country, but I had a book to finish; a book which I had actually started before the war and in which I wanted to explain atoms to the layman. I worked on it solidly from morning to night, and it appeared as *Meet the Atom*, but was no great success. Since then I have written several more books on similar subjects, none of them a best-seller. I enjoy writing and always get a little bit of fan mail which makes me feel good, and I tell myself that if just a few youngsters are attracted to physics by one of my books and become good scientists then it was well worth writing.

At the end of those two weeks, just at I thought I could now begin to enjoy California, I got a phone call from John Cockcroft, head of the atomic laboratory at Chalk River, Canada. He said he would like me to take on the job of running a Division in the new Atomic Energy Research Establishment (A.E.R.E.) which was being started at Harwell in England (not far from Oxford) and wanted me to come to New York and meet my prospective deputy, a Dr Coburn. Could I be in New York the next evening? Well, in these days of planes, everything is possible. With a bit of telephoning I got a reservation on a plane leaving Los Angeles next

43. Robert Cockburn (pronounced Coburn), inventor of important radar applications; then the author's deputy at the Atomic Energy Research Establishment, Harwell; later director of the Royal Aircraft Establishment, Farnborough. Knighted 1960.

morning; that meant hurried packing, leaving at dawn and catching a flight at seven from San Francisco. So I was indeed in New York twenty-four hours after I had received the phone call. At the hotel where I was to meet my future deputy the porter said 'We have no Dr Coburn'. Looking down his list he added 'There is a Dr Cockburn'. 'No,' I said, 'the name I was given was Coburn.' Eventually we tumbled to it: the porter didn't know (and neither did I) that the name Cockburn is traditionally pronounced Coburn. Later when I was in Harwell we were joined by another scientist with the German name Fuchs; the spelling sometimes caused people to pronounce it in a somewhat embarrassing way. I don't know who invented the little rhyme that explained the pronunciation of these two difficult names. It went like this: 'Said Dr Cockburn to Dr Fuchs: Where do you buy your library books? Said Dr Fuchs to Dr Cockburn: I go to a little shop in Holborn.'

Captivated by Robert Cockburn's vigorous and extrovert manner, I took to him at once, and we became good friends when we started to work together in Harwell. But Cockcroft suggested I should first spend a few weeks in Canada, at Chalk River where most of the Canadian atomic energy effort was concentrated. It was too short a time to find my way around or to form lasting friendships. But it was exciting to see my friend Kowarski in charge of the construction of a heavy-water reactor; later he used the experience he had gained in building the first reactor in France (and indeed in Europe).

Only one funny story comes to mind. One young physicist, convinced that nobody ever read his technical reports, proceeded to test his conviction by persuading a typist to insert in his report a few lines in the vein of '...seven green fairies were dancing in the moonlight...'. The test proved him wrong: he received a courteous enquiry from Cockcroft concerning the relevance of green fairies to the subject of his report. And one limerick may deserve quoting. When to the horror of the radiation protection squad it was found that a filing cabinet next to a secretary's desk contained quite a strong radium source, Nicholas Kemmer (later Professor at Edinburgh University) wrote:

> A typist, proficient in Morse,
> Sat for weeks on a radium source
> Until a pink rash
> . — . . —
> The rest of the story is coarse.

As I left Canada I was pursued by a phone call from a big American film company who had previously written to Lise Meitner asking her for advice on a film about the atomic bomb and requesting her permission to be represented by an actress. Lise Meitner had flatly refused, declaring that to see herself in a film would be as bad as having to walk the length of Broadway in the nude. As to getting the details right, she referred the film makers to me. The phone call missed me because I had boarded a train to Halifax, from where I was to embark. The train journey – thirty-six hours – was a bit of a nightmare at first because as soon as the train had started moving I discovered that I had no cash on me, and for a while I starved; eventually I found someone who knew me and helped me out.

Soon after my arrival in England the film company got hold of me; I got a letter from a lawyer who invited me to come and for a suitable fee give permission to be represented by an actor, and also to give my opinion on the script they had prepared or rather on the scene in which I was to appear. My visit to the lawyer introduced me to a man who had achieved my ultimate ambition; he had a grand piano in his office! He showed me that script, and I could see at a glance that it was a caricature of what really happened in laboratories, and that the people concerned, such as Lise Meitner and Otto Hahn, were ridiculously out of character. I said I would think about it, took it home and decided that I might as well have some fun by writing an alternative script.

A few days later I took it to the lawyer. He asked me for my fee, and I asked for ten pounds. He said that was too little (I wish I had asked for a hundred) and gave me twenty for the script and for my permission to be represented. I didn't care; if Oppenheimer and other participants in the atomic bomb work were willing I didn't mind, and if they weren't the film wouldn't be made. Actually it did in the end reach the cinemas but showed only Americans and gave the impression that no other country had participated in either the research or the production of the A-bomb. All the scientific details were ludicrously out of true, and the whole rather pretentious production was soon forgotten.

I had been interviewed in Canada for the British Civil Service and had been offered the post of Division Leader in the Atomic Energy Research Establishment which was being built at Harwell, with the grade of Deputy Chief Scientific Officer. At about the same time I received the O.B.E., abbreviation for Officer of the Order of the British Empire, as I explained to my father. He was very proud and went around telling people that I had become the deputy chief scientific officer of the British Empire; tongue in cheek, but I dare say some foreigners believed him. The post was roughly equivalent to a professorship; I took to travelling first class on trains! There were no working facilities at Harwell as yet, where

44. The author about 1945.

an old military airfield was being converted for the purpose, and a couple of months was spent in an office in London, at the top of the Shell-Mex House on the Strand, with very little to do. I dare say I might have done more had I had more initiative. As it was I spent my time working out the mathematics of fluctuations in chain reacting systems, the most ambitious project in mathematical physics I have ever tackled, and one which later got published with so many mistakes that I don't dare look at it and certainly never tell anybody where to find it.

During my stay in London I lived at the flat of Lotte Meitner-Graf, a distinguished photographer, the wife of Walter Meitner, my mother's youngest brother; he was an industrial chemist with a job in Manchester whereas his wife had her studio in London. It was a pleasant life and I got rather spoiled there, and in addition I had frequent occasion to visit her at the studio and admire her work. She really was a wonderful artist; many of her photographs, for instance of Jawaharlal Nehru, Bertrand Russell and Yehudi Menuhin are found on book or gramophone jackets, and a poster with her photograph of Albert Schweitzer was seen all over England for a while, urging young people to do voluntary work in developing countries. She rather specialized in portraits of musicians and physicists; there of course I was sometimes able to help her make contact.

When Harwell became ready for use I went to live there, staying first in the guest house; later, on announcing that I was about to get married, I was assigned a prefabricated house made from aluminium and glass-fibre insulation. The subsequent winter was the most severe in my experience; I placed tumblers containing some water in the various rooms, where they stayed frozen solid for weeks on end. Only one room I kept heated. On one occasion, coming back from a weekend and having switched off the heating as I left, I found that the water supply had frozen right below the main stopcock, which of course I had turned off to prevent the pipes from bursting. I had to fill my electric kettle with snow and melt it repeatedly until I had enough hot water to trickle onto the water pipes, which eventually began to gurgle and give. My engagement was soon broken off, but I was allowed to keep the prefab until I left in the autumn of 1947.

There was a lot of building still going on and when the thaw came the mud was indescribable. I got myself a pair of wellingtons which I wore from morning to night and which earned me the nickname 'Commissar Frisch'. But on the whole I spent my time in my office continuing my calculations on the fluctuation of chain reactors and trying to develop methods for carrying out precise measurements in spite of those fluctuations, which of course were a source of error. Occasionally I apologized to my deputy (now Sir Robert Cockburn) that I let him do all the work of hiring people and organizing them into units. But I think he enjoyed that; at any rate he reassured me that if I had two good ideas a year I had earned my salary.

Cockburn had the gift of strategy which I was lacking. When a problem was put to him he would immediately try to estimate

45. Posing for the press about 1946, from left to right: William Penney, the author, Rudolf Peierls, John Cockcroft.

how many and what kind of people were needed to tackle it, whereas my first impulse was always to tackle it single-handed, which of course was foolish. But Cockburn appreciated my own gifts, and I remember – and I feel this was one of the greatest compliments ever paid to me – how once he said 'When I have been knocking my head against a brick wall all day, along comes old Frisch and sniffs around and finds a back-door.'

In those days there was still a shortage of some foods and consequently a black market (about which I knew nothing). One day Cockburn invited me to share a chicken that his wife had managed to acquire by means which were not made clear. We sat down at the table with our mouths watering; and then we found that the chicken was totally inedible! I had never tried to chew anything as tough and rubbery. Cockburn nearly blew a fuse. I said it must have been a fighting cock.

When I first came to Harwell I found that the Ministry of Works had installed mahogany benches in all the laboratories, so I borrowed a hand drill and made several useless holes, just to set a precedent and encourage people to use these elegant surfaces as laboratory benches, for screwing things down when needed. From time to time I wandered over to look at the work that was going on in the reactor division where the first nuclear reactor in Britain was being built, a pile of graphite blocks without any cooling which – like Fermi's first pile – could only be run at very low power but was very useful for a variety of measurements. Its name was ZEEP, zero energy experimental pile. The next reactor, with water cooling and capable of much higher power, became named BEPO, for British Experimental Pile, the 'o' being added just to make it sound better. The man in charge was Charles Watson-Munroe, with such an Australian accent that I never knew whether he was saying pile or power; the two words sounded completely alike to my ear.

I was still staying in the bachelors' quarters, Icknield House, when one evening I was startled to hear very loud music out of my fire place. I had already gone to bed, but there was no thought of sleep with that noise; so I stormed upstairs in my dressing gown and knocked at the door of the room above. There was Ken Smith, a young physicist, who had placed his loudspeaker in front of his fireplace, using the wall of his room as a baffle board; our fireplaces were served by the same flue, so of course about half of the music emerged into my room! He was very apologetic and instantly turned the volume down; we soon became good friends. In fact he later decided to continue his study of physics by working for a doctorate in Cambridge, and I became his supervisor. Soon afterwards he became head of the physics department of the University of Sussex.

It was also in Icknield House that I tried to get better acquainted with Klaus Fuchs, whom I had briefly met in Los Alamos. He was a quiet and retiring person; I tried to persuade him to make music with me, having discovered that he played the violin, but he said he was out of practice, and we never got together. I recall a discussion among several of us on what to do with torn socks and no wives to mend them. (Socks were made from wool in those days and didn't last long.) One could buy adhesive patches to stick over the holes, as I had discovered, and by now my socks consisted largely of such patches. Klaus Fuchs, when asked what he did with his torn socks, replied simply 'I wear them'. I greatly respected him, not only for his skill in mathematical physics but also for the humane way in which he looked after his people; he was head of the division for theoretical physics and saw to it that canteen and living facilities were constantly reviewed and improved. He lived extremely modestly and I remember how on one occasion he offered a lift to Lise Meitner, who had come to visit Harwell and whose official car had for some reason been delayed. He drove her to the station hell for leather, at a full thirty miles per hour; it was all his old jalopy could do, and she missed the train. At that time none of us had any idea of the sensation which Fuchs would create a few years later.

Let me jump ahead. In 1950 I had moved to Cambridge, when I got a phone call from the B.B.C. asking me to give, at very short notice, a brief talk about the hydrogen bomb. At first I was puzzled why they wanted a talk on the chemical concept of the 'hydrogen bond' about which I knew next to nothing. When that error was cleared up I was glad to accept the invitation; but I had to work very fast. Moreover it was necessary to obtain clearance for a talk on such a highly secret piece of weaponry; I had to make sure that my script didn't give away any secrets. I phoned my old friend Peierls, but he was busy and suggested that Klaus Fuchs might be able to give me clearance. When I phoned Fuchs he merely said that it was not convenient just then. So I got in touch with another man who I knew could give clearance, Michael Perrin of the Civil Service; he agreed that if I came to his office just before recording my talk he would read my script and tell me if anything ought to be deleted.

Most of that talk I wrote on my portable typewriter in the train to London, and from the station I went straight to Perrin's office. There the first person I met was Eric Welsh, who had done intelligence work in Norway and come back with a Norwegian wife. He received me with unusual solemnity, saying it was absolutely essential that I have clearance for my talk; I was puzzled why he should sound so serious. I had to wait a while before I could speak to Perrin. He went through my manuscript quite fast, suggested one or two small changes and then let me go off to the B.B.C. My talk was duly recorded, to be broadcast after the news that same evening. Then I met my fiancée, who had come to the B.B.C. to have lunch with me and who surprised me by saying that one of my buddies – she thought the name was Fuchs – had been arrested for treason. I was staggered; 'Treason? Nonsense!' I said 'It couldn't even be speeding because his car couldn't do that; perhaps he had parked on the wrong side of the road.' 'No' she insisted, 'it was something like high treason.'

When I found by listening to the next news bulletin that indeed he had been before Bow Street magistrates accused of giving away secrets to Russia, I thought at first, this must be some sort of frame-up. I was dumbfounded when eventually I heard that he had admitted it, and that in his naive way he assumed it was a minor offence and that after a rap over the knuckles he would be back in Harwell because his work was important. He certainly didn't dream that he was going to get fourteen years. I still believe that Fuchs acted entirely out of sincere motives. He was the son of a German clergyman and brought up to act according to his conscience. He felt that Communism was the nearest approach to Christianity to be found in today's world and that it was not right that our brave allies should be kept in ignorance of the advances in our weaponry. After his release (having served about nine years) he went to East Germany and got a Professorship; but he found that there it was not so easy to combine mathematical work with the welfare of the people who worked with him. He was told to leave welfare to those who were in charge of it.

But now I must go back to the spring of 1947 when, quite unexpectedly, I was offered the Jacksonian Professorship at the

University of Cambridge. It now seems very odd that I hesitated for several days. Partly it was just laziness; after nearly two years in the Civil Service, with Robert Cockburn doing most of my work, I was tempted to carry on that easy life. Yet it was a very distinguished chair I was being offered; it had been held by Sir James Dewar (inventor of the vacuum flask, among other things), Sir Edward Appleton (father of ionospheric research) and by Lord Rutherford. I must say, the thought of stepping into Rutherford's shoes terrified me; I would rattle around in any one of them!

Sir John Cockcroft had succeeded him in 1938 but had been absent from Cambridge during the war when he played an important part in organizing research, first on radar, then on the A-bomb. Now he was director at Harwell, and it was his old chair I had been offered; but when I asked his advice he refused to influence me in either direction. Did he want to keep me in Harwell? Or would he like to replace me by somebody more efficient, but did not want to say so? I shall never know. My Swiss friend, Egon Bretscher, with whom I had climbed mesas in Los Alamos, became my successor.

The rumour had got around. The Oxford physicist Lord Cherwell said 'what is this Frisch I hear? You are going to the Eastern Zone?' But Chadwick, whose advice I sought, said, only a fool would refuse; so I accepted.

Cambridge 1947– . . .

While I was spending a brief summer holiday with my proud parents in Sweden I got a letter from Sir Lawrence Bragg; he wanted to propose me for a Fellowship in Trinity College and asked me for a *curriculum vitae*. My high-spirited reply quoted as one of my achievements 'designing parts for the Copenhagen cyclotron, which later had to be replaced at great cost' and as my main hobby 'lying in the sun and doing nothing'. Much later I heard that copies of that letter, circulated by Bragg among the Fellows, had helped in getting me elected.

College life suited me well: the company of articulate and well-informed men, a roomy apartment overlooking a spacious courtyard right next to Wren's library, and all my needs attended to. I am still a Fellow for life but no longer live in College; married men don't as a rule. But I still have a meal there most days and feel part of that intellectual ambience.

It was in London, through my photographer-aunt Lotte Meitner-Graf, that I met Ulla Blau, a graphic artist born, like myself, in Vienna; and in the spring of 1951 we married. She opened my eyes to a world of visual art to which I had paid little attention. We had a daughter and a son, in quick succession, both now working elsewhere but still in touch with us. My son has even paid me the compliment of studying physics; so now we can have fun talking

46. Author and his wife
Ulla, newly married in 1951.

shop. My daughter, with a degree in social sciences, enjoys improving the lot of others; we could do with more people like that.

I have given my reasons why I don't want to say much about the twenty-five years I spent at the Cavendish Laboratory, but I shall touch briefly on the highlights as I see them. Others may see them elsewhere, and I apologize to all those to whose work I have not done justice. But first I want to say a few more words about my parents.

In March 1938 when Hitler annexed Austria my father was still in Vienna. He escaped such humiliations as having to scrub the street under Nazi supervision; he had no hooked nose and was not instantly spotted as a Jew. But in November the German Ambassador in Paris was killed by a desperate young German Jew – the incident on which Michael Tippett based his oratorio *A*

47. The author with his two children, Monica and Tony, about 1965.

Child of Our Time – and a night of terror followed, remembered as '*Kristallnacht*' because the streets were aglitter with crystals of broken glass from the smashed windows of Jewish shops; a great many Jews were rounded up and taken to concentration camps.

The first I knew about it was when my mother phoned from Vienna and shouted desperately that father had been taken away by the S.S. (which stands for 'Schutzstaffel', Himmler's black-shirts). The next two months are a confused nightmare in my memory. Niels Bohr sent one of his scientists to Germany to talk to a high-ranking official who was the father of one of Bohr's friends; but I believe nothing came of that. However, my father's boss, Dr Bermann, had managed to escape to Sweden before the Kristallnacht and to restart his business; he offered father his old job back if he could come. On the strength of that offer a high Swedish official, Justizieråd Alexandersson, promised that my father would get a labour permit, should he reach Sweden. I don't know how much that was Bohr's influence.

Non-political prisoners in concentration camps were often released if they had somewhere to emigrate to; of course they were relieved of any significant assets. My parents were comparatively lucky; they brought most of their furniture and a good many books, as well as the fine Schiedmayer grand piano, a couple of years older than myself, which I still play.

Even in the frightful conditions of a concentration camp my father showed his qualities, advising and cheering many of his fellow prisoners. One of them said one evening he had had enough and would go over the fence that night; a euphemism for suicide since anybody approaching the fence was of course shot by the guards. Father looked thoughtful, pretended to remember something, and said 'do you know, last night I had a dream; I dreamt you had been released!' The man hesitated and did not 'go over the fence'; the next day he was released. My father never believed in dreams; he had made up the story to delay the desperate man and give him one more day, with its small chance of freedom. That his prophecy had by luck come true gave father considerable prestige and enabled him to put heart into others.

When he arrived in Sweden with my mother he seemed little changed, only a bit thinner. For a fortnight or so he was busy with the typewriter, and I saw some of what he wrote about his experiences; pretty gruesome reading in places. When he had finished he told me it was all in that envelope (which he sealed) and that he would never talk about the concentration camp again. And he never did.

Soon after she came to Stockholm my mother was knocked down

48. Joseph S. Mitchell, Regius Professor of Physic at the University of Cambridge. A physician who uses physics in his unremitting fight against cancer. (Sketch by the author.)

by an errand boy on a tricycle; the fall broke her hip, which was pinned but never healed. She was in pain for the rest of her life. In 1948 father was pensioned off at the age of almost seventy, and my parents came to join me in Cambridge. A growth on his jaw, which had regressed under radiation treatment in Stockholm, was now recognized as a secondary growth from lung cancer. The 'old man's friend' – pneumonia – might have shortened his suffering but was chased away with antibiotics. Soon afterwards father was admitted to Addenbrooke's Hospital. I shall be forever grateful to Joseph Mitchell, a physicist turned physician, who looked after him for several weeks during his last illness, brought him cheer and saved him much suffering.

Father kept his clear mind and his sense of humour until the day of his death. His firm belief that there was no life after death was a source of strength to him; he felt that the party was over and that it was time to go. The last entry in his diary, a few days before he died, was 'I owe the barber one shilling' followed by 'Diary discontinued'. I tried to find the barber, but there was a new one who didn't know who had been before him.

Mother lived another two years, in a nursing home for old people. I had made her a wheelchair, but she never learned to manipulate the wheels and mostly just sat. She had a fairly large room, with space for her grand piano, and it gave her pleasure to have me play to her when she could no longer play. But once she did play: I persuaded her to visit me in College where four undergraduates were recruited to carry her, wheelchair and all, two floors up the steep stairs to my palatial set of rooms. A glass of wine persuaded her to try my Bechstein upright, and though her fingers were weak and unsure she managed to raise a ghost of her former charm and sparkle.

When I told her one day that I was about to get married she lamented that now she had lost her only son, and seemed puzzled and unconvinced when I assured her that, on the contrary, she had gained a daughter; though she took to my wife, and affection grew between them. But she got steadily weaker, and a few months later she died in her sleep.

Nine years later Lise Meitner retired from her research in Sweden and bought a flat in Cambridge which we had helped her find; still full of vigour at eighty-one, giving lectures abroad and walking on her beloved Austrian mountains; at eighty-seven she recovered from a heart attack, and a year later from a fall that broke her hip. She still felt pride in receiving a share of the Fermi Prize (together with Hahn and Strassmann) in belated recognition of her part in the discovery of uranium fission. But then her strength was failing slowly, and in the end she slipped away almost imperceptibly, a few days before her ninetieth birthday, having outlived all her brothers and sisters. She was buried in a country churchyard in southern England, next to her youngest brother.

Now back to the Cavendish Laboratory. Lord Rutherford, after his sudden death in 1937, had been succeeded by (Sir) Lawrence Bragg, famous son of a famous father. He was particularly interested in studying the structure of crystals through the diffraction of X-rays, following the discovery in 1912 of X-ray diffraction by Max von Laue. When he came to Cambridge some such work had already been started by John D. Bernal, who had

49. Lawrence William Bragg (1891–1971). Originator of crystal analysis by X-rays. Head of the Cavendish Laboratory, Cambridge, from 1938 to 1952. Knighted in 1941.

boldly tackled complex molecules of living matter such as proteins. The first dramatic success came when John Kendrew and Max Perutz unravelled the structures of the oxygen-carrier proteins, myoglobin and haemoglobin, which give the red colour to meat and blood; the first crude model of a myoglobin molecule looked like a plumber's nightmare to me! But within a few years the arrangement of the thousands of atoms in those complex molecules had been analysed in detail.

At about the same time, Francis Crick and James Watson (a visitor from the U.S.A.) came to understand the double helix of the genes, those twisted scrolls on which the entire specifications for a mouse or a man are copied into every cell in a code, using four letters in groups of three, now deciphered. That was perhaps the greatest single advance in biology since Darwin's concept of natural selection, and it happened right under my nose; but I didn't understand what was going on.

The analysis of those complex molecules needed very complex calculations which would have been impossible without the electronic computers. I often wandered across the courtyard to the

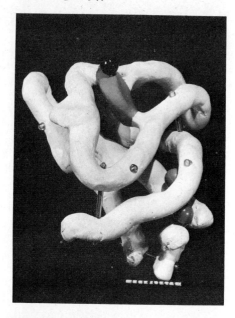

50. The first crude model showing the winding and branching chain of protein in the myoglobin molecule, resulting from the early work of John Kendrew (1917). Knighted in 1974.

51. Max Perutz, Austrian-born scientist, assembling a detailed model of the haemoglobin molecule. The vertical bars support the 'atoms'. Joint Nobel Prize with John Kendrew in 1962. Companion of Honour, 1976.

mathematical laboratory where Maurice Wilkes was directing the construction of one of the first electronic computers in this country, a roomful of several thousand radio valves mounted on upright panels and connected by wires running in all directions as if a host of spiders had been busy. I tried to understand how it worked and was thrilled to learn about tricks that were new to me, such as the use of the same group of valves for storing either a number or an instruction. One day, by chance, I met John von Neumann on the London train and started to tell him about our EDSAC (Electronic Digital Sequential Automatic Calculator) and its special advantages. He cut me short: 'Listen, Otto, those tricks are all known. The EDSAC has *one* advantage: it works!' I was a bit deflated but also cheered to hear from the master that we were not doing so badly in Cambridge.

Today a machine of comparable power goes in one's pocket. Although we hear and talk a lot about computers, I am convinced that most of us are nowhere near to understanding their full importance. Two hundred years from now historians will say that the computer changed our world much as the steam engine did two hundred years previously, if not more so.

Another event which depended on computers and which I was able to watch was the birth and growth of an entirely new discipline, radioastronomy. Like the time-of-flight method I mentioned earlier it is a spin-off from radar. Observers had noticed that a radar signal contained more noise when the aerial was pointed at the sun, and other points in the sky also emitted weak irregular radio waves though nothing much could be seen there with an optical telescope.

An ordinary radar 'dish' (a concave reflector some 10 feet in diameter) was neither sensitive enough, nor did it resolve fine detail; progress was made in two ways. At Jodrell Bank near Manchester a huge radar dish (250 feet in diameter) was built through the initiative of (Sir) Bernard Lovell; it was a versatile and useful instrument which could be pointed at any spot in the sky. In Cambridge a cheaper but subtler method was developed

52. Martin Ryle, whose initiative and inventiveness contributed greatly to the creation of the new science of radioastronomy in the 1950s. Knighted in 1966, shared the 1975 Nobel prize with Anthony Hewish. (Photo Walter Leigh.)

through the initiative and ingenuity of Martin Ryle. Soon after the war he started with a few small radar dishes linked together to achieve the resolving power of a much larger one. With that radio telescope an area of the sky could be mapped: the rotation of the earth was used to scan the sky, and measurements over many days were combined mathematically, with the help of the new electronic computers. Later the system was improved so that a small spot on the sky can be tracked and its radio intensity mapped in very fine detail; that improved technique is now increasingly used all over the world.

Successive surveys of the radio sky have revealed a multitude of novel objects in the sky, such as jets a million light-years in length, and fast-spinning stars as heavy as the sun but only a few miles in size; a whole new world of which the optical telescopes had shown us nothing. But most immediately amazing is the incredible sensitivity of those huge installations. When the Mullard Radio Astronomy Observatory was inaugurated here in 1958 each guest at the luncheon found a little white card by his plate, with these words on the back: 'In turning over this card you have used more energy than all the radio telescopes have ever received from outer space.'

53. The Cambridge 5-km radio telescope commissioned in 1972. By combining signals of all eight dishes in a computer an angular resolution (i.e. detail of mapping the sky) equivalent to the use of a dish 5 km in diameter can be obtained. It is the first radio telescope to achieve a mapping detail better than large optical telescopes installed on good mountain sites.

Martin Ryle's ingenuity and his skill in organizing such an enterprise was rewarded with a knighthood in 1966, and he shared the 1974 Nobel Prize with Anthony Hewish, a member of his team who had recognized, in 1967, the significance of uncannily regular pulses from one radio source. For a few weeks they had kept that observation a close secret, fearing the storm of sensational headlines (EXTRATERRESTRIAL INTELLIGENT LIFE DISCOVERED BY

CAMBRIDGE SCIENTIST!!!) if the press got hold of it. Now those 'pulsars' (a few dozen more have been found) are regarded as collapsed stars made largely of neutrons; if you like, gigantic nuclei some miles in size, held together by gravity, spinning rapidly and sweeping a beam of radio waves around the sky several times a second. Many features of those cosmic lighthouses have now been well accounted for.

It is a splendid demonstration of the unity of physics that study of certain rare stars should let us learn more about matter as dense as atomic nuclei; our radio telescopes, looking out into space, are supplementing the giant accelerators, microscopes that let us look into the inside of nuclei.

And that brings us back to nuclear physics, the subject of the department which I was supposed to direct when I was called to Cambridge. There was good work going on, and I followed my nature by keeping a low profile. My only venture worth recalling from that time was my suggestion to a new research student, to measure the spin and magnetic moment of some radioactive nuclei by the use of molecular beams, with which I had become familiar in Hamburg. But the student never needed my advice. He was Ken Smith, whose loudspeaker had startled me in Harwell when loud music bellowed from my fireplace. In 1952 he succeeded in measuring the spin of bismuth-210 (known from the old days as radium-E), together with Henry Bellamy; and when we went to the physics conference at Glasgow our result (obtained at the last minute) was received with joy and announced with a fanfare by Chien Shiung Wu (nicknamed 'Miss triple U' by her many friends) who found it helped her with the theory of the beta disintegration of radium-E. Both Bellamy and Smith became professors soon afterwards, and the use of molecular beams was not continued at the Cavendish Lab.

But at that conference I heard the first report about a possible new research tool, destined to replace Wilson's cloud chambers which at that time were being built with thicker and thicker walls

54. Early picture of a bubble chamber traversed by many fast electrons, entering through the window below and deflected by a magnetic field. The faster they are the less curved are their tracks; some are seen to spiral as they get slower. Through the top glass plate the bottom plate (both circular) can be seen, as well as the bolts that connect them.

and filled with gas at ever higher pressures in order to increase the chance that a high-energy particle should hit a nucleus in the chamber, forming secondary particles whose tracks were worth studying. Those monstrous chambers with inch-thick walls reminded one of the great lizards of the jurassic age, finally superseded by more versatile creatures like mammals and birds.

That new creature, the bubble chamber, was invented by Don Glaser who thought that instead of observing tracks of droplets in a supercooled gas (as C. T. R. Wilson had done forty years earlier)

215

one might observe tracks of bubbles in a superheated liquid. His first photograph, shown in Glasgow, was pretty crude, just a row of a dozen bubbles in a thimble-size chamber. It looked a bit like the rows of bubbles you see in a glass of beer (which had not been his inspiration, or so he told us); but it showed that the principle was sound.

Three years later I met Don Glaser again at the university at Ann Arbor (near Detroit); we made music (he plays the viola) and talked about bubble chambers, which had reached a size of some inches, with bigger ones being built. While cloud chambers needed minutes to recover after each expansion, bubble chambers could be operated at one photo a second or faster, well matched to the accelerators which had passed the 1000-MeV mark and delivered short bursts of fast particles every few seconds. Moreover each photo would be likely to show a collision, with secondary tracks demanding measurement which traditionally took an hour or so. Glaser stressed that there would be a severe bottleneck at that point unless fast semi-automatic measurement methods were quickly developed.

That was a challenge after my own heart. Back in Cambridge, one phone call secured permission to use a grant for developing a track-measuring device instead of improving a cyclotron. A young engineering graduate, Alan Oxley, designed and built the electronics, and within a year we had a prototype, good enough for useful research. In Switzerland, California and elsewhere similar devices were built. The film with the track images is projected on a screen with a cross in the middle, and the operator moves the film carriage so that several successive points on each track coincide with the cross, each time pressing a button which causes the position of the carriage to be coded on tape. The use of two or more films, showing the bubble chamber from different directions, provides three-dimensional information when the tape is later taken to a computer; the program makes it reconstruct the tracks in space, work out their angles and curvatures, and even try various suggested interpretations of what happened in each collision.

Later we tried various ways of improving the speed and accuracy; if an expert is one who has made every conceivable mistake I became very expert in this field. One excessively ambitious machine was scrapped on the day it was completed; we worked out that it would have taken too long to write the required complex computer programs. Moreover while it was being built I had thought of a better machine, more automatic and much faster. For that machine to come off – as it did – needed a lot of luck.

The first piece of luck was that we were refused a government grant which would have compelled us to drop our development and buy a ready-made machine. Was I happy when that refusal came! A small laser, recently acquired for the practical class, solved two of my problems, even though it often wouldn't start until I had touched it repeatedly with a comb, electrified by combing my hair. Today you switch those little lasers on and off like desk lamps, and they have become cheaper, smaller and much more powerful.

It was also lucky that John Rushbrooke, the senior man in my group, believed in my idea and persuaded our boss, Professor Nevill Mott (knighted in 1962 and recipient of the 1977 Nobel Prize for his work on the structure of solid matter) to dip into one of his special funds and buy a small computer to act as the brain of my machine. 'Small' meant the size of a wardrobe and costing over £10000. We found a young mathematician, Julian Davies, who wrote the required programs, correcting and greatly expanding the rough one I had written while on holiday in Devon.

One big piece of luck was the arrival of another research student, Graham Street, who at once showed uncommon inventiveness. Without his great skill and insight, his untiring hard work and his unfailing enthusiasm my complicated scheme would probably not have come to anything.

The final bit of luck was the intervention of my old friend Lew Kowarski, on whose initiative a quite different kind of track-measuring machine had been developed at CERN in Geneva, but who liked mine and arranged for the next International Conference

55. The author and his machine 'SWEEPNIK' for measuring bubble-chamber tracks, at the Cavendish Laboratory, Cambridge, in 1971.

on Nuclear Measuring Techniques to be held in Cambridge in 1970. High-energy physicists came from all over the world and climbed all over our SWEEPNIK (the nickname for our machine, sweeping up information like a sputnik); we had got it to work during the night before the conference! Helsinki and Honolulu were interested in buying one; John Rushbrooke made the right contacts so that we were offered capital to set up a factory; as the oldest I was made chairman and still am.

On the whole, my health is still good. During a severe winter at Cornell University (upstate New York) I got an ear infection that made me partly deaf; but it also damaged my equilibrium organ just enough so that I no longer get seasick. Cataracts in both eyes were successfully operated, and with my contact lenses I see quite well. It was probably exposure to neutrons during my A-bomb work that affected my eyes; some may think I had deserved worse punishment.

Quite a few of the students whose research I had supervised have now moved into academic posts, and not the least of my pleasures in travelling is to meet them and to see what they are doing now, in Switzerland, Israel or the U.S.A.; a good number of them turned up to celebrate my seventieth birthday (but that is where we came in). My mother, in wise foresight of the Cambridge regulations, saw to it that I was born on the first day of October; that gave me an extra year before I had to retire, at the end of the academic year in which – on its first day! – I reached the age of sixty-seven. And retiring in a leap year gave me an extra day!

All my life I have been interested in the design of scientific devices, even more than in the results which I or others might obtain with their help. The firm we started for manufacturing SWEEPNIK is still growing and giving me scope to do what I like best. I can do much of my work at home and interrupt it to play the piano whenever I like. I am a lucky man.

Further reading

M. Born, *My Life and Views* (Charles Scribner's Sons, 1968).

P. C. W. Davies, *Space and Time in the Modern Universe* (Cambridge University Press, 1977).

O. R. Frisch, *The Nature of Matter* (Thames and Hudson, 1972).

B. Hoffman, *Albert Einstein, Creator and Rebel* (Hart-Davis, Mac-Gibbon, 1972).

K. Mendelssohn, *The World of Walther Nernst* (Macmillan, 1973).

S. Rozental (ed.), *Niels Bohr: His Life and Work* (North-Holland Publishing Co., 1967).

V. F. Weisskopf, *Knowledge and Wonder* (Heinemann Educational, 1964).

Acknowledgements

Reminiscences come mainly out of the author's memory. On the other hand, the people who have helped me by supplementing and correcting my hazy recollections are so numerous that I could not list them even if I remembered them all. To mention merely a few would be invidious; I just tender my thanks to all of them.

Particular thanks are due to Giles de la Mare and to Charles Lang, who both gave me valuable advice on how to improve and expand my original draft; and to Mrs Judy Lowe who cheerfully typed and retyped my messy manuscript until it was fit for the printer.

The author and publisher would like to thank the following for permission to reproduce illustrations.
2: Photograph by Lotte Meitner-Graf. 5: The Cavendish Laboratory, Cambridge. 6: AIP Niels Bohr Library, Bainbridge Collection. 9: Photograph by Lotte Meitner-Graf. 13: Copyright Photopress, Zurich. 17: The Cavendish Laboratory, Cambridge. 18: Photograph by Lotte Meitner-Graf. 22: Harold Agnew. 23: Photograph by Lotte Meitner-Graf. 24: The Niels Bohr Institute, Copenhagen. 28: Hanne Zapp-Berghäuser, Fotografin. 29: Dennis Bracher. 30: United Kingdom Atomic Energy Authority. 33: Los Alamos Scientific Laboratory. 34: Photograph by N. Metropolis. 36: Photo CERN. 38: Drawing by Martyl. Courtesy of the *Bulletin of the Atomic Scientists*. Copyright 1957 by the Educational

Index

Index

Index

226